T0074518

An Interactive Guide to Quantum Optics

Online at: https://doi.org/10.1088/978-0-7503-2628-5

IOP Series in Quantum Technology

Series Editor: **Barry Garraway** (School of Mathematical and Physical Sciences, University of Sussex, UK)

About the series

The IOP Series in Quantum Technology is dedicated to bringing together the most up to date texts and reference books from across the emerging field of quantum science and its technological applications. Prepared by leading experts, the series is intended for graduate students and researchers either already working in or intending to enter the field. The series seeks (but is not restricted to) publications in the following topics:

- Quantum biology
- Quantum communication
- Quantum computation
- Quantum control
- Quantum cryptography
- Quantum engineering
- Quantum machine learning and intelligence
- Quantum materials
- Quantum metrology
- Quantum optics
- Quantum sensing
- Quantum simulation
- Quantum software, algorithms and code
- Quantum thermodynamics
- Hybrid quantum systems

A full list of titles published in this series can be found here: https://iopscience.iop.org/bookListInfo/iop-series-in-quantum-technology.

An Interactive Guide to Quantum Optics

Nikola Šibalić
Quantum Computing Center, Quantum Machines, Tel Aviv, Israel

C Stuart Adams
Department of Physics, Durham University, Durham, UK

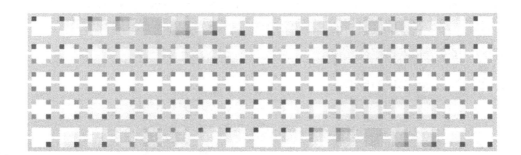

IOP Publishing, Bristol, UK

Nikola Šibalić and C Stuart Adams have asserted their right to be identified as the authors of this work in accordance with sections 77 and 78 of the Copyright, Designs and Patents Act 1988.

ISBN   978-0-7503-2628-5 (ebook)
ISBN   978-0-7503-2626-1 (print)
ISBN   978-0-7503-2629-2 (myPrint)
ISBN   978-0-7503-2627-8 (mobi)

DOI   10.1088/978-0-7503-2628-5

Version: 20240101

IOP ebooks

British Library Cataloguing-in-Publication Data: A catalogue record for this book is available from the British Library.

Published by IOP Publishing, wholly owned by The Institute of Physics, London

IOP Publishing, No.2 The Distillery, Glassfields, Avon Street, Bristol, BS2 0GR, UK

US Office: IOP Publishing, Inc., 190 North Independence Mall West, Suite 601, Philadelphia, PA 19106, USA

We dedicate this work to those without the privilege to do science. We hope for a future where everyone has the opportunity to explore and develop their curiosity.

Contents

Author biographies		**x**
1	**Introduction**	**1-1**
	References	1-5
2	**Discovery of quantum optics**	**2-1**
2.1	Optics timeline: from corpuscles and waves to photons	2-2
2.2	Classical, semi-classical or quantum?	2-21
2.3	From radio frequency electromagnetic fields controlling quantum systems to quantum optics and back	2-21
	References	2-25
3	**What is quantum in quantum optics?**	**3-1**
3.1	Einstein's introduction of the photon concept	3-1
3.2	New types of dynamics brought by quantum theory	3-4
3.3	When are phenomena or technology quantum?	3-5
3.4	Single quanta, optics and biological systems	3-6
	References	3-8
Part I	**One quanta**	
4	**One quanta**	**4-1**
4.1	Introduction	4-1
4.2	What is a photon?	4-2
4.3	Photon properties	4-3
4.4	Photons as qubits	4-5
4.5	Photon polarization	4-7
4.6	The Bloch and Poincaré sphere	4-9
4.7	Photon polarization: linear basis	4-13
4.8	Density matrix	4-14
4.9	A photon beam splitter	4-15
4.10	Rotations	4-16
4.11	Waveguide beam splitter	4-17
4.12	Photon creation: a single-photon emitter	4-21
4.13	Mathematical description of light–matter interactions	4-22
4.14	Two-level system driven by a near-resonant field	4-23

4.15	Decay and decoherence	4-24
4.16	Detecting single photons: homodyne detection	4-30
	References	4-34

5   Measurements: projective and non-destructive   **5-1**

5.1	Bang, and what happens next: quantum jumps and quantum regression theorem	5-2
5.2	Effects of jumps	5-6
	5.2.1 Post-jump transients viewed in frequency: Mollow triplet	5-7
	5.2.2 Post-jump transients viewed in detection: photon anti-bunching	5-11
	5.2.3 Jumps communicating information overlap: entanglement and interference	5-11
	5.2.4 Quantum non-demolition measurements	5-22
5.3	Effects of information leakage: new coherent dynamics	5-24
5.4	On phase in quantum physics	5-28
	5.4.1 Quantum dynamics in full colour: a new convention for figure making	5-28
	5.4.2 Measurement induced feedback with no decay leading to fixed relative phase: **electromagnetically induced transparency** transient example	5-34
	5.4.3 Well defined relative phase, with no global phase: example of two inverted two-level systems	5-37
	References	5-39

6   Fighting environmental noise and imperfections   **6-1**

6.1	Ramsey sequence	6-2
6.2	Dynamical decoupling (bang–bang control)	6-5
6.3	Spin-echo	6-7
6.4	Switching external perturbation: Dicke narrowing	6-8
6.5	Non-Markov environments: coupling initially independent field modes	6-10
	References	6-14

Part II   Two or more quanta

7   Two photons   **7-1**

7.1	Introduction	7-1
7.2	Two-photon interference	7-2
7.3	Hong–Ou–Mandel effect: Fock basis	7-2
7.4	Bell states	7-7

7.5	Polarization-entangled photons	7-7
7.6	Bell's inequality	7-9
7.7	Two-qubit visualization: rotations	7-11
7.8	Linear optics quantum computing	7-13
7.9	Hong–Ou–Mandel effect: non-symmetric network	7-13
7.10	Controlled-NOT gate	7-15
	References	7-20

8   Seeing entanglement: counting and correlation   **8-1**

8.1	Introduction	8-1
8.2	Quantum erasure	8-2
8.3	Photon experiments	8-3
8.4	A quantum circuit model	8-6
	References	8-7

9   Strong interactions   **9-1**

9.1	Introduction	9-1
9.2	Jaynes–Cummings model	9-2
	9.2.1 Exotic multi-quanta states	9-3
	9.2.2 Squeezed light	9-5
	9.2.3 Cat states	9-7
	References	9-9

Part III Outlook

10   Outlook   **10-1**

10.1	Semiotics	10-2
10.2	Finding and connecting insights	10-3
10.3	Sharing in accessible and reusable form	10-4
10.4	Science versus decision making	10-5

Appendix A   **A-1**

Author biographies

Nikola Šibalić

Nikola Šibalić studied physics at the University of Belgrade, Serbia and received a PhD from Durham University, UK, before doing postdoctoral research in Denmark and France. Currently he dedicates his days working in the quantum computing industry as product manager and solution architect, and his nights to developing and maintaining an ever growing array of open-source tools for knowledge sharing and (re)use. He relaxes by practising classical mechanics in the form of swing dancing. Find out more at nikolasibalic.github.io.

C Stuart Adams

C Stuart Adams studied physics at the University of Oxford and received a PhD on laser physics from the University of Strathclyde in Glasgow, UK. He completed postdoctoral work in Germany and the United States before starting a research group at Durham University in 1995. He was awarded the Thomson medal in 2014 by the Institute of Physics (IoP) and the Holweck Prize in 2020 by the French Physical Society and IoP for pioneering work in atomic physics and quantum optics. He is co-author of the textbook *Optics f2f*. Find out more at etotheipiequals.github.io and opticsf2f. github.io/Opticsf2f_CodeBook/

Together they co-authored previously *Rydberg Physics*, IOPP (2018)

IOP Publishing

An Interactive Guide to Quantum Optics

Nikola Šibalić and C Stuart Adams

Chapter 1

Introduction

The fact that we live at the bottom of a deep gravity well, on the surface of a gas-covered planet going around a nuclear fireball ninety million miles away and think this to be normal is obviously some indication of how skewed our perspective tends to be.

Douglas Adams (1952–2001)

All great civilizations in human history were interested in light, and made various attempts to control and understand it (figure 1.1). In 17th century Europe, two ideas about nature of light existed: Christiaan Huygens (1629–95) was advocating the wave nature of light, while Isaac Newton (1642–1727) saw light as a set of small particles (corpuscle hypothesis). Later, in the 19th century, the wave nature of light became dominant. Thomas Young (1773–1829) performed an experiment with two apertures and showed interference of light, which was previously seen only for waves. James Clerk Maxwell (1831–79) explained electromagnetic waves as coupled oscillations of electric and magnetic fields. It seemed that the wave interpretation of light could explain everything. Yet seeing electromagnetic waves as something discrete, as small indivisible parcels of energy, was revisited first in Max Planck's (1858–1947) description of black-body thermal radiation spectrum, and then Albert Einstein's (1879–1955) description of photoelectric effect.

Initially, the light quanta concept met significant resistance[1]. Particularly because our ability to experiment with individual quanta developed later. In 1952 Schrödinger wrote [2][2].

---

[1] See for example [1].
[2] Later, we shall come back to the question that Schrödinger was asking in 1952, *Are There Quantum Jumps?*

doi:10.1088/978-0-7503-2628-5ch1

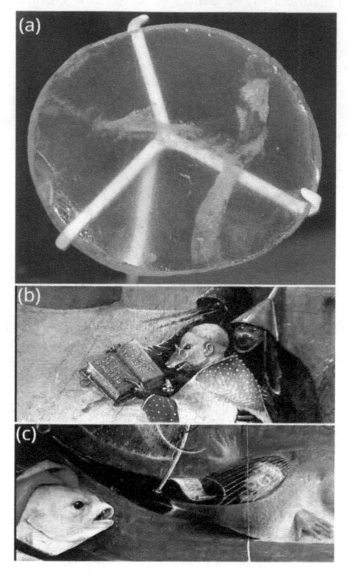

Figure 1.1. The Nimrud lens, plano-convex rock crystal from Assyria made in 750BC–710BC, was possibly used as an early lens (a). Euclid's and Ptolemey's early theories on light were explored and expanded in works of Islamic scholars like Ibn al-Haytham *Book of Optics* (1011–1021), influencing early European researchers. The first wearable eyeglasses were thought to be invented in 1284, in Florence by Salvino D'Armate. By the time of Hieronymus Bosch (1450–1516) optical glasses were popular even among monsters (b and c) in religious paintings. Picture credits: (a) Wikimedia Commons, CC-BY-SA (b and c) Hieronymus Bosch, Triptych of Temptation of St Anthony (1505–06), detail, Museu Nacional de Arte Antiga, Lisbon.

we never experiment with just one electron or atom or (small) molecule. In thought-experiments we sometimes assume that we do; this invariably entails ridiculous consequences ...it is fair to state that we are not experimenting with single particles, any more than we can raise Ichthyosauria in the zoo.

But Schrödinger's statement that *we never experiment with just one* is no longer true. From the 1970s and 1980s onwards we began to experiment with single particles, first photons then atoms, ions, molecules and electrons. Subsequently, experiments on single quanta became commonplace, so much so, that Schrödinger's *ridiculous consequences* have became a part of the lived experience of many physicists, and are increasingly finding use in a diverse range of applications.

A modern viewpoint could be that quantumness requires at least two key ingredients, counting and interference[3]. We count particles (quanta), and everything in the quantum world is described by a wave. As waves interfere, when and where we count depends on interference. Central to this idea is wave–particle duality. The essence of quantumness to be both wave and particle at the same time. The essence of our classical world is that we can only observe either wave or particle-like property at any one time.

As quantum theory developed it became increasingly hard to find analogies in our everyday world. Two million years of human evolution on the savanna did not prepare us for our astonishing journey into the quantum world. As Richard Dawkins said [3], 'modern physics teaches us that there is more to truth than meets the eye; or that meets the all too limited human mind, evolved as it was to cope with medium-sized objects moving at medium speeds through medium distances in Africa'. Understanding phenomena at high speeds, astronomical distances, the extremely slow pace of geological change, the long-time averaged trends of climate change, and Nature at very small scales requires quite some stretching of the human imagination. Einstein himself said that 'common sense is actually nothing more than a deposit of prejudices laid down in the mind prior to the age of eighteen'. If the speed of light was 10 km h^{-1}, relativistic effects would be understood as something intuitive, self-evident—just like Mr Tompkins in George Gamow's book (figure 1.2) —and we might not have needed to wait for Einstein to come along and make a revolution in our understanding of space-time.

Today, we know that a property may only exist if we prepare a system in a state with specific information about that property. If the world has no information about a photon in a box, we can only speculate about the existence of that photon. A photon in the double-slit experiment does not pass through either slit in the sense of a billiard ball—it is everywhere and can see both open slits. The beauty of quantum mechanics is that it can tell us, when we can correlate something in our system with the position of the photon, what is the probability of observing the photon position in some given space interval. Similarly, we cannot find an analogy for the electron— it is neither wave nor particle—these labels make sense only in our macroscopic world.

Another pioneer of quantum physics, Richard Feynman, in his famous Messenger Lecture series at Cornell University in 1964, said that [4]

---

[3] There is no universally agreed definition of quantum versus classical. We can observe, quantumness in experiments that do involve counting, such as the homodyne measurements on single photons discussed in section 4.16. Also, as waves and particles can be classical it is also argued that more esoteric concepts such as entanglement are the essential difference of the quantum world (see section 3.3).

Figure 1.2. Georgiy Antonovich Gamov (1904–69) gave a theoretical explanation of alpha decay by quantum tunneling, invented the liquid drop model and the first mathematical model of the atomic nucleus, and worked on radioactive decay, star formation, stellar nucleosynthesis and Big Bang nucleosynthesis. In his popular book *Mr Tompkins in the Wonderland* he imagines the world in which speed of light is only 10 km h^{-1}. Reprinted with permission from *Mr Tompkins in Paperback*, George Gamow, John Hookham. Copyright (1993) Cambridge University Press 1965.

I think I can safely say that nobody understands quantum mechanics.

Reflecting on quantum weirdness and the differences between the quantum world and everyday experience Feynman added

> *The difficulty really is psychological [...] the perpetual torment that results from you saying to yourself, 'But how can it be like that?' [...] Don't keep saying to yourself [...], but how can it be like that? Because you'll get down a drain, you'll get down into a blind alley from which nobody has yet escaped. Nobody knows how it can be like that.* Full lecture: 7:08–8:40.

But, photons and electrons are useful concepts—the indivisible building blocks of modernity. We cannot represent a photon as a crowded carriage of classical physics (figure 1.3): no amount of gears, whistles, water waves, balls and ropes[4] will recreate dynamics of quantum objects. If one cannot hope to construct a classical analogy of a photon, can we try instead to create correct visualization of their *states and dynamics*. This is what we attempt in this book, taking a tour of a gallery of useful

---

[4] Formally called in this context 'hidden variables'.

Figure 1.3. Phenomena in quantum mechanics cannot be understood as limit of our macroscopic everyday observations purely by changing physical constants. This is because observations in quantum mechanics (see chapter 5) have a much more profound effect on dynamics than that is the case in classical and even relativistic physics. Cartoon by Nick D Kim, scienceandink.com. Used by permission.

conceptual pictures used in understanding quantum optics phenomena. As in quantum physics we are all still students, all exhibits are interactive. On our tour we shall survey what we can say and cannot say about these bizarre quantum objects. And once we have established why photons are different we shall talk about how their unusual aspects form the basis for new quantum technologies.

Before getting into quantum, in chapter 2 we shall begin with some history of how we ended up here.

References

[1] Lamb W E Jr 1995 Anti-photon *Appl. Phys.* B **60** 77–84
[2] Schrödinger E 1952 Are there quantum jumps? Part II *Br. J. Philos. Sci.* **3** 233–42
[3] Dawkins R 2004 *A Devil's Chaplain: Reflections on Hope, Lies, Science, and Love* (Boston, MA: Mariner Books)
[4] Feynman R 1964 The Messenger Lectures, Cornell Full lecture: 7:55 https://www.youtube.com/watch?v=w3ZRLllWgHI

IOP Publishing

An Interactive Guide to Quantum Optics

Nikola Šibalić and C Stuart Adams

Chapter 2

Discovery of quantum optics

The real voyage of discovery consists not in seeking new landscapes, but in having new eyes.

Marcel Proust (1871–1922)

The important thing in science is not so much to obtain new facts as to discover new ways of thinking about them.

William Lawrence Bragg (1890–1971)

How did we end up here? The origins of quantum optics begin with our observations of the Sun. Scientists in the nineteenth century accumulated data on the solar spectrum, see figure 2.1, that led directly to the birth of quantum mechanics and hence quantum optics.

 LabBricks literature graph for this chapter can be found at https://labbricks.com/#9/bpR6KJdvQd

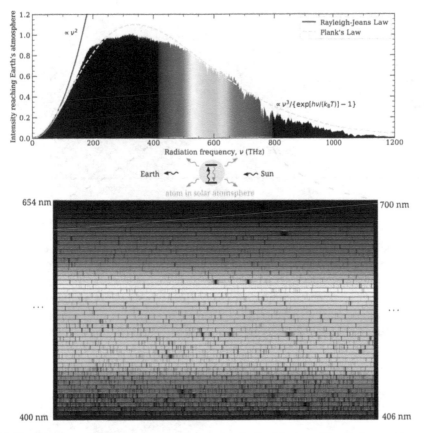

Figure 2.1. Sunshine is quantum! Solar radiation that reaches the Earth's atmosphere (top) cannot be explained with classical Rayleigh-Jeans Law, and requires the introduction's of Planck's quantization factor h. It is then well explained as black-body radiation spectra. Slight large-scale discrepancy is due to Sun not being a simple black-body emitter. More interesting is the fine structure of the spectrum that shows sharp dark lines (bottom) from elements present in the Sun's atmosphere that scatter away absorbed radiation. To explain this discrete spectra, quantum theory of atomic and molecular structure needed to be developed. Data sources: ASTM E-490 solar spectral irradiance from NREL (top); Solar Flux Atlas from 296 to 1300 nm, Robert L Kurucz, Ingemar Furenlid, James Brault, and Larry Testerman: National Solar Observatory Atlas No. 1, June 1984 (bottom), Credits: N.A. Sharp/KPNO/NOIRLab/NSO/NSF/AURA (CC BY 4.0).

With hindsight we might guess that it was the observation of discrete lines made in the solar spectrum that led to the ideas of quantization, whereas, in fact, it was the thermodynamics of light that led Planck and Einstein to introduce the quantization of light, see section 3.1. The quantization of the energy levels of electrons in matter followed shortly later.

2.1 Optics timeline: from corpuscles and waves to photons

This section is available as an interactive figure at http://doi.org/10.1088/978-0-7503-2628-5.

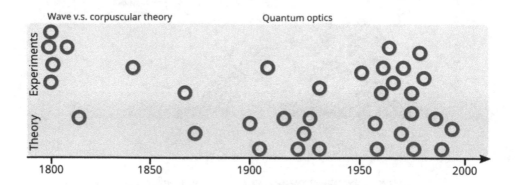

1800—Frederick William Herschel, discovery of infra-red

Frederick William Herschel (1738–1822) was working in Bath as an astronomer when he started to measure the heating effect of different colors of light coming from the Sun, resolved through the refraction in a prism. He was using his thermometer and was measuring relative heating of a thermometer placed in different colors of the solar spectrum. When the thermometer was just beyond the red end of the spectrum, Herschel noticed that relative temperature of the thermometer was highest, indicating strong heating. He concluded that some invisible rays, just beyond the red part of the spectrum, are coming from the Sun and causing heating.

Image credits: reproduced from [1].

1801—Young's double-slit experiment

The experiment of Thomas Young (1773–1829), English physician and physicist, established strong evidence for wave theory of light at the time when the majority of scientists thought that Newton's corpuscular theory was sufficient. Young used wave theory of light to explain color of thin films and he calculated wavelengths of the principal colours in the visible spectrum. Some years later, in 1817, Young proposed that light waves are transverse, not longitudinal as it was assumed, explaining polarization of light.

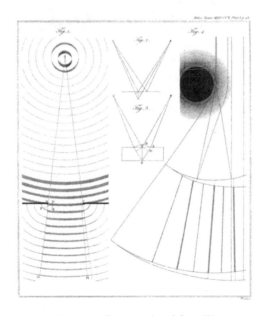

Image credits: reproduced from [2].

1801—Johann Wilhelm Ritter, discovery of ultraviolet

Following on Herschel's discovery of infrared radiation beyond the red part of the spectrum, Johann Wilhelm Ritter (1776–1810) was examining if there is light beyond the blue part of the spectrum. He used silver chloride, which is a chemical compound that decomposes when exposed to light, leaving darker silver visible. Ritter noticed that silver chloride doesn't change in color in the red part of the spectrum, but becomes darker in the blue part of the spectrum. He then put the silver chloride beyond the blue part of the spectrum, where no light was visible, and discovered that darkening of silver chloride is even faster there. He concluded that there is additional invisible light—now called ultraviolet—in this part of the spectrum that was causing the reaction.

Plants can have ultraviolet patterns. These can be seen by bees, as well as some other insects, birds and reptiles. In the picture we see side by side a Mimulus flower photographed in visible light (left) and ultraviolet light (right). The dark strip visible only in UV is the so-called *nectar guide* for bees.

Credit: This Mimulus nectar guide UV VIS.jpg image has been obtained by the authors from the Wikimedia website where it was made available by User:Plantsurfer under a CC BY-SA 3.0 licence. It is included within this article on that basis. It is attributed to User:Plantsurfer.

1802—Discovery of absorption lines in the solar spectra

Joseph von Fraunhofer (1787–1826) started as a glassmaker, and developed new types of glasses, standardized glass production methods, and thanks to this created scientific instruments with unmatched quality. He also used diamonds to create optical diffraction gratings with grooves spaced only 0.003 mm apart. Using spectrometers he built, he mapped more than 500 dark lines in the solar spectrum, that now we understand originate from selective absorption of the Sun's radiation by elements present in the solar atmosphere.

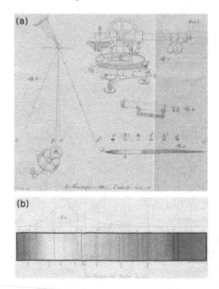

(a) The spectroscope developed by Fraunhofer; (b) The spectrum of sunlight drawn by Fraunhofer. Image credits: VIII. Bestimmung des Brechungs- und Farbenzerstreuungs-Vermögens verschiedener Glasarten, in Bezug auf die Vervollkommnung achromatischer Fernröhre. München: [Verlag nicht ermittelbar], 1817. ETH-Bibliothek Zürich, Rar 5099, Public Domain.

1809—Etienne-Louis Malus discovers polarization

Back in 1669, Danish physician, mathematician, and physicist, Erasmus Bartholin (1625–98) discovered double refraction in Icelandic spar (calcite). A light beam passing through the crystal would be split into two, and upon rotation of the crystal, one of the beams would would remain stationary while the other would rotate with the crystal. These two beams were called *ordinary* and *extraordinary*, respectively, but the phenomenon was not understood at the time.

Using two calcite crystals, with crystal principal axis set at variable angles, French physicist Étienne-Louis Malus (1775–1812) observed the effect of the second crystal on ordinary and extraordinary rays transmitted through the first calcite. He deduced that the calcite changed some property of the light. By looking through a crystal of icelandic spar at the sunset reflected in a window of the Palais du Luxembourg in Paris he found that the same property of light was changed also upon reflection from surfaces. Following corpuscular theory, he explained this by particles of light having two poles—hence introducing the concept of polarization of light for the first time.

Erasmus Bartholin, Experimenta crystalli Islandici disdiaclastici quibus mira & insolita refractio detegitur, (Copenhagen, 1669) E-L Malus, Sur une Proprietede la Lumiere Reflechie, *Memoires de Physique et de Chimie de la Societe d'Arcueil* II, 143 (1809, Paris, Mad.Ve:Bernard, quai des Augustins) E-L Malus, Memoire sur de Nouveaux Phenomenes d'optique, *Moniteur* No. 72, 277 (1811). Image credits: 445 nm laser comes from left to calcite crystal, and shows clearly birefringence forming two rays at the output. Fluorescence within the crystal shows light paths in the crystal. Jan Pavelka, CC BY-SA 4.0, via Wikimedia Commons.

1815—Mathematical description of wave theory of light

In the period from 1815 to 1825 Augustin-Jean Fresnel (1788–1827), French civil engineer and physicist, established the mathematical basis for wave theory of light, and realized that light is polarized in transverse direction to the direction of the propagation. Fresnel made numerous contributions including birefringence, diffraction, Fresnel–Arago laws, Fresnel equations, Fresnel integrals, Fresnel lens, Fresnel number, Fresnel rhomb, Fresnel zone, Huygens–Fresnel principle, Phasor representation.

Image credits: p. 773 in Oeuvres complètes d'Augustin Fresnel. Tome 1/publiées par M M Henri de Senarmont, Emile Verdet et Léonor Fresnel (1866), showing reprint of work A Fresnel, Mémoire sur la loi des modifications que la réflexion imprime à la lumière polarisée (1823), showing derivation of Fresnel's sine law for field amplitude reflection coefficient of s-polarized light. Source gallica.bnf.fr/Bibliothéque nationale de France.

1842—Photograph of the solar spectrum

Edmond Becquerel (1820–91) photographed the solar spectrum including the UV region.

Picture credit: Edmond Becquerel, Spectres solaires, 1848, images photochromatique. Courtesy of musée Nicéphore Niépce, Ville de Chalon-sur-Saône [3].

1856—Demonstration of the absorption of heat by CO_2 and water vapor.

Eunice Newton Foote (1819–88) discovered the principal cause of global warming, by measuring heating in dry air, damp air and a CO_2-filled atmosphere. She concluded that 'An atmosphere of that gas [CO_2] would give to our Earth a high temperature; and if as some suppose, at one period of its history the air had mixed with it a larger proportion than at present, and increased temperature from its own action as well as from increased weight must have necessarily resulted.' [Eunice Foote, 'Circumstances Affecting the Heat of the Sun's Rays', American Journal of Science and Arts (1856)]. In addition to her scientific work, she was an inventor with several patents and active in the women's rights movement.

1868—Ångstrom publishes solar spectra atlas

Ångstrom published 'Recherches sur le spectre solaire', an atlas of observed solar spectral lines with wavelength given in units of 10^{-10} m. His data led first to the Rydberg formula and hence the quantum theory of atoms (Bohr model).

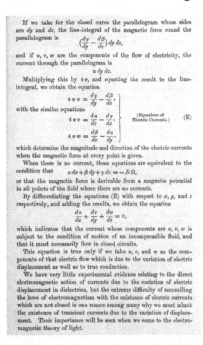

Image: reproduction from Ångström, Anders Jonas, Recherches sur le spectre solaire, (Upsal: W Schultz, 1868).

1873—J C Maxwell publishes a treatise on electricity and magnetism

Scottish physicist James Clerk Maxwell (1831–79) provided a unifying description of all observed electric and magnetic phenomena to date in his monumental work 'A treatise on electricity and magnetism'. Crucially, he introduced another source of magnetic field: the electric field change that according to his mathematical argument based on consistency of theory, should have the same effect on production of magnetic field as current. This enabled him to explain light as electromagnetic phenomena, i.e. coupled oscillations of electric and magnetic field.

Image credits: reproduction of p. 233 from James Clerk Maxwell, A treatise on electricity and magnetism, 3rd edition (Oxford: Clarendon Press, 1881) showing Ampère's circuital law, and arguing for Maxwell's addition, namely that change in electric field should be taken as a source of magnetic field just as current is taken as source of magnetic field by Ampère, and highlighting importance of this addition for electromagnetic theory of light. Source: Internet archive, digitalized collection of University of California Libraries

1900—Black-body spectra, Max Planck

In order to explain black-body spectra, Max Planck (1858–1947) introduces the idea that oscillators in the black-body can only have a discrete number of energy elements, each containing energy proportional to the frequency of oscillations, with the constant of proportionality now known as Planck's constant.

> then to the radiation present in the medium. We must now give the distribution of the energy over the separate resonators of each group, first of all the distribution of the energy E over the N resonators of frequency v. If E is considered to be a continuously divisible quantity, this distribution is possible in infinitely many ways. We consider, however—this is the most essential point of the whole calculation—E to be composed of a very definite number of equal parts and use thereto the constant of nature $h = 6.55 \times 10^{-27}$ erg sec. This constant multiplied by the common frequency v of the resonators gives us the energy element ε in erg, and dividing E by ε we get the number P of energy elements which must be divided over the N resonators. If the ratio is not an integer, we take for P an integer in the neighbourhood.
>
> It is clear that the distribution of P energy elements over N resonators can only take place in a finite, well-defined number of ways. Each of these ways of distribution we call a "complexion", using an expression introduced by Mr. Boltzmann for a similar

Image credits: reprinted excerpt from The old Quantum Theory, D Ter Haar, page 84, translation of Max Planck, On the Theory of the Energy Distribution Law of the Normal Spectrum, copyright Pergamon press (1967), originally published as Max Planck, *Verh. Dtsch. Phys. Ges. Berlin* **2**, 237 (1900).

1905—Einstein introduces the concept of the photon

In his 1905 paper, Albert Einstein (1879–1955) acknowledged the tremendous success of the wave theory of light, and carefully highlighted that it had been chiefly tested in observations of time averages rather than instantaneous values. Then he focused attention on transient phenomena involved in the transformation of light, and used probabilistic interpretation of entropy to argue for introduction of energy quanta that cannot be divided, and can be produced and absorbed only as complete units—that is a concept of the photon.

> It seems to me that the observations associated with blackbody radiation, fluorescence, the production of cathode rays by ultraviolet light, and other related phenomena connected with the emission or transformation of light are more readily understood if one assumes that the energy of light is discontinuously distributed in space. In accordance with the assumption to be considered here, the energy of a light ray spreading out from a point source is not continuously distributed over an increasing space but consists of a finite number of energy quanta which are localized at points in space, which move without dividing, and which can only be produced and absorbed as complete units.

Albert Einstein, Über einen die Erzeugung und Verwandlung des Lichtes betreffenden heuristischen Gesichtspunkt, *Ann. Physik* **17**, 132 (1905) Image credits: reprinted excerpt from A B Arons and M B Peppard, Einstein's Proposal of the Photon Concept—a translation of the Annalen der Physik paper of 1905, *American Journal of Physics* **33**, 367 (1965), with the permission of AIP Publishing.

1909—G I Taylor photographs interference pattern for weak light

G I Taylor (1886–1975), physicist and mathematician, imaged the shadow of a needle illuminated with very low light intensity, trying to observe deviation from Huygens-style wave interference when only a few light quanta are present in the wave front. For the lowest illumination, exposure time was 2000 hours or about 3 months. That means that in an apparatus there is approximately one photon per every 100 000 transit times it takes a photon to travel from source to image plate. In spite of such low intensity, a sharp interferometric pattern was obtained on the photographic plate [4].

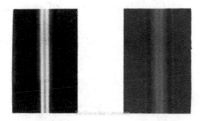

Image: G I Taylor's demonstration of diffraction bands caused by a thin wire in feeble light. Two black and white photographs of diffraction bands caused by a thin wire in strong and feeble light. The exposures were 4 minutes (left) and 400 hours (right). Although the light photons only arrived one at a time, the characteristic diffraction pattern was observed. Credits: Courtesy of and copyright the Cavendish Laboratory, University of Cambridge.

1917—Einstein's introduction of stimulated emission

Rederiving Planck's black-body radiation law from radiative equilibrium (along the lines of Boltzmann's derivation of equilibrium distribution from the collision equations), Albert Einstein (1879–1955) found, based on energy balance, that there exists a process of stimulated emission, and from momentum balance, he found that photons in stimulated emission must be emitted in the same direction as radiation causing it.

> If a ray of light causes a molecule hit by it to absorb or emit through an elementary process an amount of energy $h\nu$ in the form of radiation (induced radiation process), the momentum $h\nu/c$ is always transferred to the molecule, and in such a way that the momentum is directed along the direction of propagation of the ray if the energy is absorbed, and directed in the opposite direction, if the energy is emitted. If the molecule is subjected to the

Image credits: reprinted excerpt from The old Quantum Theory, D Ter Haar, page 182, translation of Albert Einstein, On the Quantum Theory of Radiation, copyright Pergamon Press (1967), originally published as Albert Einstein, *Physikalische Zeitschrift* **18**, 121 (1917). See also derivation from present point of view in R Friedberg, Einstein and stimulated emission: a completely corpuscular treatment of momentum balance, *American Journal of Physics* **62**, 26 (1993).

1924—Indistinguishable particles and Bose–Einstein statistics[1]

In 1924, Satyendra Nath Bose (1894–1974), Indian mathematician and physicist, rederived Planck's black-body spectrum purely from maximization of entropy, assuming quantum nature of light, and—crucially—fundamental indistinguishability of photons when counting possible micro realizations of the states. Albert Einstein then published in 1925 statistics of ideal (Bose) quantum gas, predicting Bose–Einstein condensation at low temperatures. Although indistinguishability of the particles was implicitly in statistical theory for several decades—see Gibb's paradox and N! factors—this derivation brought clear focus to the assumption that photons are fundamentally indistinguishable, and was met initially with resistance as it seemed to imply '...a mutual influence of the particles on each other of a kind which is at this time still completely mysterious.' as Einstein put it, leading to condensation and reduced pressure under dense, cold conditions of quantum gas. Yet these were real phenomena, and later Bose and Fermi statistics was incorporated in treatment of all many-body quantum problems.

---

[1] For discussion from a modern viewpoint see [5].

1927—P A M Dirac, 'The Quantum Theory of the Emission and Absorption of Radiation'

P A M Dirac (1902–84) introduced quantization of the electromagnetic field, resolving the apparent conflict between corpuscular and wave theories.

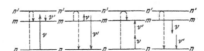

Image credits: reprinted excerpt from [6].

1931–Prediction of two-photon absorption

Maria Goeppert-Mayer (1906–72) in her doctoral thesis discussed two-photon (de)excitation. This was experimentally verified only in 1961. when new high-intensity laser light sources were invented. This phenomenon later gave rise to nonlinear microscopy. She was awarded the Nobel Prize in 1963 for her work on a mathematical model that defined the structure of an atom's nucleus. She worked on unpaid positions in academia for a long time, being granted a full-time salaried academic position only when she was 53 years old.

Image credits: reprinted figure 4-1 with permission from [7]. Copyright 1931 John Wiley & Sons.

1935—Thin coatings

Katharine Burr Blodgett (1898–1979) extended the research of Irving Langmuir (1881–1957) on single-molecule thin film creation on water surface, creating methods for depositing stacks of monomolecular coatings, allowing creation of thin film coatings for glass and metal that are of precisely controlled thickness. In this way she made a coating that made a glass more than 99% transmissive—thanks to destructive interference of reflections from air-coating and coating-glass interface —that had influence on all optical devices, from cinematography, submarine periscopes to optical assemblies in research. She also patented a color gauge for measuring thickness of the thin films.

Image credits: Kathrine Burr Blodgett at General Electric Research Laboratories, Smithsonian Institution Archives, Science Service Records, 1902–65 (Record Unit 7091). Courtesy of Smithsonian Institution Archives.

1935—EPR paradox

The Einstein, Podolsky and Rosen (EPR) paradox highlighted one key property of newly developed quantum theory: importance of measurement for establishing reality. Science up to that point had been based on the idea that the world around us has certain properties regardless whether it is observed or not. These properties one can call 'real' or as the EPR trio put it 'If, without in any way disturbing a system, we can predict with certainty (i.e. with probability equal to unity) the value of a physical quantity, then there exists an element of physical reality corresponding to this physical quantity'. However, by examining the case of measuring non-commuting variables on two entangled systems, the trio showed that reality of one part of the system seems to change without direct interaction with the other part of the system. This opened a question: can it be that somehow quantum theory can be expanded with some additional variables that determine outcome of the sequential measurement on two parts of the system without requiring this abrupt change in 'reality' of the physical property of the system. Three decades later, J S Bell in 1964 ruled out the existence of such local, hidden-variable theory, based on an experimentally measurable scheme. This firmly established that the importance of measurement and lack of 'reality' of entangled system properties before such measurements is a genuine property of Nature.

The experiment that we shall discuss here is aimed at testing whether there really is a correlation in polarization directions of the type described in the foregoing. The ideal way to test this point would be to measure the polarizations of each member of a statistical ensemble of pairs of photons produced by positron-electron annihilation; and to see whether the polarizations are always perpendicular in every system of axes, as predicted by the theory. But this is not yet possible in practice. Nevertheless, there has been done an experiment which, as we shall see, tests essentially for this point, but in a more indirect way. This experiment[10] consists in measuring the relative rate, R, of coincidences in the scattering of the two photons through some angle, θ, for the following two cases:

(1) When the planes π_1, and π_2 formed by the lines of motion of the scattered quantum and the original direction are perpendicular ($\varphi = 90°$, where φ is the angle between the planes).
(2) When the planes are parallel ($\varphi = 0$).

These cases are illustrated in Fig. 1. The photons originate at the point, 0.
In the first case, photon 1 is scattered by an electron in a block of solid matter at the point A, through some

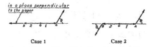

Case 1   Case 2

FIG. 1. Schematic representation of apparatus for observing correlations between photons.

Image credits: reprinted excerpt with permission from [8]. Copyright 1957 by the American Physical Society. The paper introduced two decades after the original EPR paper a more experimentally accessible version of the original gedanken experiment, this time using polarized photons.

1956—Hanbury Brown–Twiss experiment

R Hanbury Brown (1916–2002) and R Q Twiss (1920–2005) were working in the field of radio astronomy, looking at the ways to expand on the limits of usual Michelson-interferometer arrays of radio telescopes, where signals recording both phase and intensity were combined (interfered) and used to form radio telescopes with better angular resolution. They suggested first using intensity correlations between the distant dishes to measure angular size of distant objects. Since only intensity had to be recorded, this was feasible back then even if the baseline (distance between individual telescopes) was large [9]. They then applied the same idea to detection of optical radiation, applying their technique to measure the angular diameter of the Sirius star. Today their experimental configuration is often used for measuring correlations within a single beam [10].

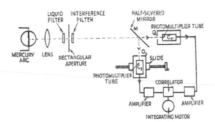

The figure shows experimental apparatus for measurement of correlations. Reproduced from [11] (figure 2) with permission from Springer Nature.

1963—The quantum theory of optical coherence

Roy J Glauber (1925–2018) introduced quantum theory of arbitrary order correlations of the light field, and defined coherent states as fields whose higher order correlation functions factorize into first order coherence functions.

(2.17) for the photon coincidence rate and thereby define a second-order correlation function,

$$G^{(2)}(\mathbf{r}_1 t_1 \mathbf{r}_2 t_2, \mathbf{r}_3 t_3 \mathbf{r}_4 t_4)$$
$$= \mathrm{tr}\{\rho E^{(-)}(\mathbf{r}_1 t_1) E^{(-)}(\mathbf{r}_2 t_2) E^{(+)}(\mathbf{r}_3 t_3) E^{(+)}(\mathbf{r}_4 t_4)\}. \quad (3.7)$$

This too is a function whose values, even at widely separated arguments, interest us.

In view of the possibility of discussing n-photon coincidence experiments for arbitrary n it is natural to define an infinite succession of correlation functions $G^{(n)}$. It is convenient in writing these to abbreviate a set of coordinates $(\mathbf{r}_j, t_j)$ by a single symbol, x_j. We then define the nth-order correlation function as

$$G^{(n)}(x_1 \cdots x_n, x_{n+1} \cdots x_{2n})$$
$$= \mathrm{tr}\{\rho E^{(-)}(x_1) \cdots E^{(-)}(x_n) E^{(+)}(x_{n+1}) \cdots E^{(+)}(x_{2n})\}. \quad (3.8)$$

Image shows a reprinted excerpt with permission from [12]. Copyright 1963 by the American Physical Society.

1964—Bell's theorem

Einstein, Podolsky and Rosen's argument from 1935 suggested that maybe additional variables should be introduced to quantum mechanics to make a theory where the state of the system at every time will be uniquely determined and dependent only on the local parameters. J S Bell (1928–90) showed that no local, hidden-variable theory can give measurement predictions consistent with all quantum-mechanical predictions for two entangled particles [13].

1966—Measurement of photon bunching in a thermal light beam

Measuring distribution of successive photon arrival times from a thermal source, B L Morgan and L Mandel were able to observe for short delay times between photon detection that there are more photon detection events than what one would expect if the photons were completely randomly distributed in time—that is photons were bunched.

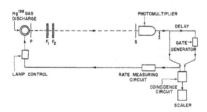

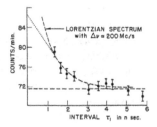

Image reprinted (figures 1 and 2) with permission from [14]. Copyright 1966 by the American Physical Society.

1967—Polarization correlation of photons emitted in an atomic cascade

The first observation of quantum entanglement with polarization states of optical photons paved a way for later tests of the EPR paradox and Bell's inequalities.

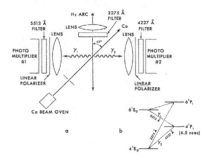

The figure shows apparatus (a) for detection of photons emitted in cascade from the Ca beam excited with light from H2 arc lamp. The level scheme (b) shows initial excitation to 6^1P_1, and marks observed deexcitation cascade $6^1S_0 \rightarrow 4^1P_1 \rightarrow 4^1S_0$. Image reprinted (figure 4-1) with permission from [15]. Copyright 1967 by the American Physical Society.

1970—Photon pairs produced by parametric down conversion

Using the setup shown in the below image, David C Burnham and Donald L Weinberg confirmed that in parametric fluorescence, initial photons from the source are converted into photon pairs that are: (i) created simultaneously; (ii) conserve total energy; and (iii) conserve total momentum. Relative polarization of the photons is fixed by the phase matching conditions on efficient down-conversion in the nonlinear crystal. This became a workforce method for production of entangled photons with solid-state components.

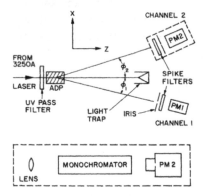

The figure shows experimental arrangement. A 325 nm laser is being converted into two spatially separated beams in phase-matched ADP (ammonium dihydrogen phosphate) crystal using the crystal's birefringence. In the horizontal plane there are two directions where phase-matched photons should go. In the channel 1 arm a narrowband band-pass filter is used in front of the photomultiplier used in counting mode. In channel 2 a monochromator is used as tunable narrowband band-pass filter. Reprinted (figure 1) with permission from [16]. Copyright 1970 by the American Physical Society.

1972—Experimental test of local hidden-variable theories

Using the polarization entangled photons from Ca cascade first measured by Carl A Kocher and Eugene D Commins in 1967, Stuart J Freedman and John F Clauser tested polarization correlations between generated photon pairs. They obtained results in agreement with quantum mechanics, and provided strong evidence against local hidden-variable theories.

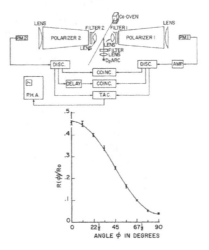

Reprinted (figures 1 and 3) with permission from [17]. Copyright 1972 by the American Physical Society.

1976—Prediction of antibunched light from resonance fluorescence

Solving dynamics for an ensemble of driven two-level atoms, one in general obtains that the ensemble dynamics will saturate at some steady state. Yet, if one looks at individual atoms, each time an atom decays emitting fluorescence, it will be starting from ground state, at that point incapable of emitting another photon. That atom will then oscillate and reach a steady state. The incapability of a single two-level atom to emit a second photon results in photon anti-bunching in resonant fluorescence, and the transient dynamics is imprinted on correlations between successive photon emissions, and it it is responsible for side peaks in the resonance fluorescence spectrum, and together with the central resonant peak it gives rise to so-called Mollow triplet.

$$G^{(2)}_{ss}(r, t_1 ; r, t_2) = \langle E^{(-)}_{ss}(r, t_1) . E^{(-)}_{ss}(r, t_2) E^{(+)}_{ss}(r, t_2) . E^{(+)}_{ss}(r, t_1) \rangle. \qquad (10)$$

For a saturated atom this is stationary and we have managed an analytical calculation in this regime beginning with the formal expression ($\tau \geqslant 0$)

$$\lim_{t \to \infty} \langle \sigma_-(t) \sigma_+(t + \tau) \sigma_-(t + \tau) \sigma_-(t) \rangle = \mathrm{Tr}_s [\sigma_+ \sigma_- \, e^{\mathcal{L}^* \tau} (\sigma_- \rho_{ss} \sigma_+)] \qquad (11)$$

where ρ_{ss} is the steady-state density operator. Generally we find

$$G^{(2)}_{ss}(\tau) = [G^{(1)}_{ss}(0)]^2 [1 - \exp(-\tfrac{3}{2} \gamma \tau) (\cosh \Omega \tau + \tfrac{3}{2} \gamma / \Omega \sinh \Omega \tau)] \qquad (12)$$

with

$$\Omega = [(\tfrac{1}{2}\gamma)^2 - 4n|\kappa|^2]^{1/2}. \qquad (13)$$

It is interesting to note that this correlation function vanishes for $\tau = 0$. This corresponds to the necessary temporal separation of photons arising from a process in which an absorption precedes every emission and constitutes a further example of the photon antibunching phenomena (see for example Stoler 1974). This arises solely in a QED treatment and if observable would provide a test for QED versus the semiclassical theories.

Reprinted excerpt with permission from [18]. Copyright 1979 by the IOP Publishing.

1977—Photon anti-bunching in resonance fluorescence

H J Kimble, M Dagenais, and L Mandel observed fluorescence of a continuously resonantly excited atom beam (using a dye laser). Analysing two-time correlation function of photon fluorescence, they observed that upon detection of one photon, during a certain time interval (determined by strength of driving necessary to excite again deexcited atoms) the probability of detecting another photon is reduced significantly below what one would expect if photons were completely uncorrelated. This so-called anti-bunching of photons is both an illustration of the quantum nature of light and an illustration of quantum jumps: while under coherent driving the atom state continuously evolves, upon detection of atom emission its state is abruptly projected on the ground state (jump down).

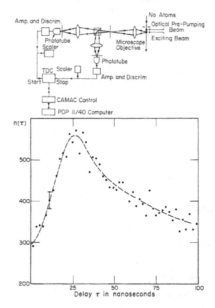

Reprinted (figures 1 and 2) with permission from [19]. Copyright 1977 by the American Physical Society.

1981—Experimental tests of realistic local theories via Bell's theorem

Using the polarization entangled photons from Ca cascade first measured by Carl. A Kocher and Eugene D Commins in 1967, Alain Aspect, Philippe Grangier, and Gerard Roger measured correlations of the photons now with improved statistics thanks to the improved efficiency of the source (two-photon laser excited), strongly ruling out local realistic theories.

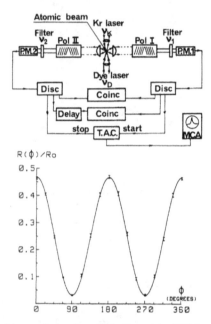

Reprinted (figures 2 and 4) with permission from [20]. Copyright 1981 by the American Physical Society.

1981—Proposal for using squeezed light to improve interferometer sensitivity

With discreteness of photons what came also is inevitable counting noise, that in principle can limit sensitivity precise optical measurement schemes. Coherent states, such as those produced by single-mode lasers, represent the minimum uncertainty states, yet their uncertainty area is equally distributed in amplitude and phase. While the area of this uncertainty 'island' cannot be reduced, when using optical non-linearities it can be shaped such that it is squeezed for example along the amplitude axis, reducing the amplitude uncertainty, or similarly phase uncertainty if the state is squeezed along the orthogonal axis. Such states were proposed by Carlton Caves to improve sensitivity of interferometers used for gravity wave detection These squeezed states are used precisely in this way in the currently operating gravity interferometers, improving their sensitivity, or equivalently, allowing detection of signals from a larger volume of the Universe surrounding the Earth.

Early discussion of minimum uncertainty states and their unitary equivalence with coherent states is presented in [21].

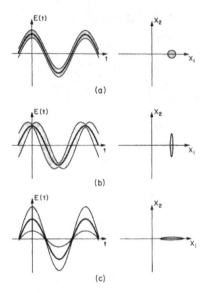

Left panels show electric field versus time for (a) coherent state, and two different squeezed states (b and c). Shading represents uncertainty in the electric field. Right panels show corresponding 'error-boxes' in the complex-amplitude plane. Reprinted (figure 2) with permission from [22]. Copyright 1981 by the American Physical Society.

1982—No-cloning theorem

This is formal proof that one cannot construct a protocol to make a copy of an unknown quantum state. Originally created as a rebuttal of some of the faster-than-light communication proposals, no-cloning theorem is the underlying element that makes current quantum communication protocols for quantum cryptography safe [23, 24].

1985—Chirped pulse amplification

Intensity of pulsed lasers was limited due to the onset of self-focusing in amplification of ultra-short pulses. Donna Strickland and Gerard Mourou found a solution for this, taking inspiration from microwave radar circuits: they stretched a short pulse through a positively dispersive medium (single-mode optical fiber), amplified low-intensity stretched pulse, before compressing it back using a negatively dispersive element, using a double diffraction grating setup. For this discovery that allowed a new frontier in research exploiting high-intensity ultra-short pulses, they shared a Nobel Prize in Physics in 2018.

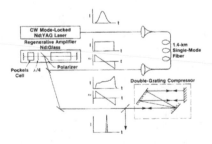

The figure shows setup for streaching, amplifying and compressing the pulse. Reprinted (figure 1) from [25], copyright (1985) with permission from Elsevier.

1987—Hong–Ou–Mandel interference

If two exactly identical photons (frequency, pulse width), whose modes are exactly spatially overlapped at a 50:50 beam splitter, arrive at exactly the same time, then completely destructive interference in amplitudes between both photons being reflected and both photons being transmitted means that at the two outputs of the beam splitter the photons will always be found in pairs. In other words, the coincidence counts of two photons, arriving in two different beam splitter outputs, will drop to zero. This was observed for the first time by C K Hong, Z Y Ou, and L Mandel, and this effect, called after their surnames Hong–Ou–Mandel interference, can be used for measuring very small offsets between the photons, either in time, or in their properties. This is often used for example in characterizing nowadays how identical are 'identical' single photons produced from single-photon sources.

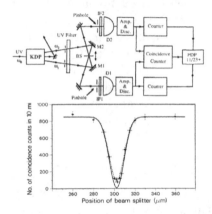

Reprinted (figures 1 and 2) with permission from [26]. Copyright 1987 by the American Physical Society.

1993—Theory of cascaded quantum systems

In two individual works, Howard Carmichael and Crispin Gardiner independently presented a theory that accounts for driving quantum systems with non-classical light of quantum sources [27, 28].

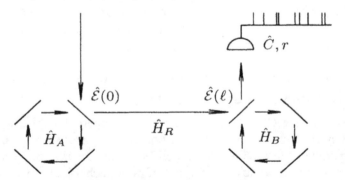

The figure shows open quantum system B cascaded with a quantum source A. Reprinted figure 1 with permission from [27]. Copyright 1993 by the American Physical Society.

1996—Quantum error correction

Quantum operations are not perfect—dynamics is continous and every pulse is usually ever so slightly off because of various reasons. As in analog commuincations, these continuous errors accumulate. In classical computers—in particular in communications and memory storage—errors can occur too, even if we discarded slow degradation of analog signals used in the past by switching to discrete states. Yet for these errors, there exist classical error-correcting codes, that spot and fix common errors. Dynamics of quantum operations, being continuous, looks more like analog problem, and poses the question whether any quantum-correction code can be applied. Even more so: can such an error-correction code be efficient enough, such that it does not kill the speed-up promised by quantum algorithms? The first affirmative answer to this was published in [29].

2001—Is entanglement needed for quantum computation?

What is the mechanism that allows quantum algorithms to achieve speed-up? Is it purely that the possible state space of few physical qubits is huge? Or do they require also entanglement in the states of the qubits to perform faster? This question arose initially with theadvancement of NMR quantum computers that could perform high-fidelity single-qubit gates, but had a hard time maintaining multi-qubit entanglement. The papers of (i) Noah Linden and Sandu Popescu [30] and (ii) Richard Jozsa and Noah Linden [31] point to the conclusion that entanglement is necessary for quantum algorithms to achieve exponential speed-up.

2.2 Classical, semi-classical or quantum?

History of optics is a story of progressively better, more-encompassing theoretical approaches. Different phenomena were selected over time to test if a phenomenon is explained with Newton's corpuscular/geometric optics or Huygens's wave optics (scalar or vector) is needed. Finally, with the development of quantum optics the question is often if observations can be explained with semi-classical theory (matter is quantized but light not) or if a full quantum optics approach (including field quantization) is needed. Depending on theory, different phenomena were the focus of analysis, from Poisson spot, to directionality and phase locking in lasing, quantum beats, photoelectric effect [32], photon counting statistics with entangled photon pairs (section 7.5), and more. Following some of these discussions can be a good exercise for checking understanding of different theories[2].

2.3 From radio frequency electromagnetic fields controlling quantum systems to quantum optics and back

Long before we mastered technology to produce coherent light, highly coherent radio frequency electromagnetic waves had been used. Ever since Heinrich Hertz's (1857–94) experiments in 1886 demonstrated Maxwell's ideas about electromagnetic waves, there has been growing development of radio technology driven by the desire for better communications. Later, during the second world war, development of

---

[2] See for example [33].

radar additionally expanded technology in the microwave region. The first coherent control of a quantum degree of freedom was established precisely by a radio frequency electromagnetic field, and from this research we inherited many methods and terms we use nowadays in optical research. Here is a short list:

- **Heterodyne** and **homodyne** detection, closely related also to **lock-in amplifiers**, was a signal processing technique developed for first radio receivers by Reginald Fessenden (1866–1932). It mixes weak signal E_s with local oscillator E_{LO}—either by combining them on a nonlinear element like a diode for radio waves, or using a square-law detector (signal $\propto$|Amplitude|^2) like a photon counter—to measure $|E_s + E_{LO}|$. Interference term $E_s E_{LO}$ provides amplification of the low signal proportionally to the amplitude of the local oscillator. Additionally, it allows frequency down-mixing which, combined with a narrowband low-pass filter, provides excellent frequency selectivity that isolates the signal from broadband noise. Both factors allow for very precise measurement of weak amplitudes, as well as phase-sensitive measurements. Using linear elements in combination with square detectors is often used in optical state reconstruction (see section 4.16).

- The first controlled driving between two discrete internal levels was achieved by Isidor Isaac Rabi (1898–88) and his team in 1938. He used an oscillating radio frequency field at 3.5 MHz to flip nuclear magnetic moments in the molecular beam of lithium chloride, whose internal state he could measure by modified Stern–Gerlach apparatus. Today, the oscillations of state between two coherently coupled levels we call the **Rabi oscillations**.

- Nuclear magnetic resonance in solid materials was first discovered by Felix Bloch (1905–83) and Edward Mills Purcell (1912–97) in water and paraffin, respectively. Bloch wrote his equation for the evolution of the magnetization **M** under oscillating field **H** in [34]. These equations explicitly look like the equation for precession of a spinning top $\frac{d\mathbf{M}}{dt} \propto \mathbf{M} \times \mathbf{H}$, and later exact geometrical representation for a 1/2 system was pointed out by [35]. That representation we now call **Bloch sphere** and use for *single*-qubit representation (see chapter 4.6). To characterize the influence of environment on the dynamics of nuclear spins, in addition to the externally applied driving pulses, T_1 **relaxation time** was introduced as characteristic time for reduction of magnetization due to thermalization of *individual* nuclear magnetic moments, that is under dissipation. T_2 **relaxation time** was introduced as a characteristic timescale for reduction of *collective* magnetization of an ensemble of nuclear magnetic moments, due to variation of local field experienced by the moments in the inhomogenous sample—that is, under inhomohenious dephasing[3]. When environment is not changing over long enough time scales (long enough self-correlation time) to allow multiple pulses that will flip magnetic moments, it is possible to recover signal using so-called **spin echo**

---

[3] Two-dimensional grey-scale images of T_1 and T_2 times are often used in magnetic resonance imaging, that maps these times for hydrogen atoms in the body, allowing differentiation between different tissues.

techniques, first introduced by [36]. This led to development of other denoising techniques for reduction of effects of inhomogenious environments. See chapter 6.

- **Purcell effect** was first discussed in [37] as a way to enhance spontaneous emission rate for nuclear magnetic moment transitions by engineering electromagnetic modes of the emitters by enclosing them in a cavity. This idea has been since used for both suppressing emission, and enhancing emission in photonic crystals, quantum dots etc, see section 6.5.

- The most common example of interference measurements, the double-slit experiment, measures second-order correlation function $\langle E^{\dagger}(t)E(t)\rangle$ of the field $E(t)$. The idea to measure higher order correlation functions, like intensity–intensity correlations $\langle E^{\dagger}(t_2)E^{\dagger}(t_1)E(t_1)E(t_2)\rangle$ came from radio astronomy, as a new type of interferometer. See the Hanbury Brown–Twiss experiment in the timeline, section 2.1. **Higher-order correlation functions** of the electromagnetic field were crucial in understanding the quantum nature of the electromagnetic field, see chapter 8.

- The first **coherent spectroscopy** was done by Norman F Ramsey (1915–2011) and I I Rabi. Previously, spectroscopy was usually done with incoherent optical sources, and for development of coherent radiation at light frequencies one had to wait for the laser, but standard radio oscillators were automatically coherent. Later, Ramsey was trying to improve on sharpness of the spectra by using slower molecular beams, allowing longer interaction times. However, making the magnetic field uniform over the long molecular beam tubes was the challenge that was limiting resolution of observed spectra. Ramsey then took inspiration from the Michelson stellar interferometer: resolution of optical telescopes is limited by the size of the aperture, and in fact, one can remove the middle part of the mirror and leave just the edges, and obtain a sharper image (at the cost of reducing total signal). Quality of the surface in between the two edges was not important, so this led Ramsey to search for an analogous setup[4]. He invented Separated Oscillatory Field Method, where a molecular beam would be interrogated with oscillatory fields only at two points, that can be placed far away, while it will freely evolve in between. Keeping magnetic field homogeneous only at these two points was technically feasible, and large separation between two interaction points allowed improved precision. This is now called **Ramsey sequence** (see section 6.1) and is used extensively for most precise measurements, from atomic clocks, to quantification of coherence of quantum systems.

- **Amplification by stimulated emission of radiation** was first demonstrated in the microwave region with masers that are still used as ultra-low noise amplifiers for Deep Space Network [38], and with feedback driving it into oscillatory mode, as for example the hydrogen maser, that can provide stable frequency

---

[4] For more details, see the interview of Norman Ramsey by Ursula Pavlish on 2006 December 4, [39].

reference. Later, the concept was extended to infrared and visible range, where it revolutionized optics, providing the first coherent sources.

- **Amplification and compression of chirped pulses.** As we have seen in the timeline, high-intensity optical pulses are amplified in the optical domain in an analogous way to radar pulses: with amplification medium being limited in the maximmum intensity it can sustain, they are expanded in time domain, amplified at weaker intensity, and then compressed using materials with different dispersion characteristics.

- **Phased arrays of coherent emitters.** Arrays of atoms, emitting light with well defined relative phases, are recently attracting attention for the possibility to shape optical pulses. Yet the same concept has been explored for long time in phased array radars, that can scan the sky simply by adjusting phase of their multiple emitters, and similar techniques are used in 5G networks antennas to direct signal more efficiently to cellular network users.

Thanks to the coherence of the ordinary radio oscillators, it was possible to do all sorts of quantum qbit manipulations long before discovery of lasers finally provided equivalent coherent sources in the optical range. Yet the first development of quantum optics was in the optical and not in microwave/radio regime. This is mainly due to the energy scales: with microwave photons having some 5 orders of magnitude smaller energies than optical photons their detection is hard. Black-body background noise at room temperature is higher in the microwave region and it is much harder to create single-microwave photon sensors. In optical domain, avalanches triggered by the photo-electron expelled with incoming radiation were enough to provide sizable electric signals for detection and good quantum efficiency[5]. And from the computational side, early quantum computers based on nuclear magnetic moments could not generate and maintain entanglement well.

All of this changes with development of superconducting cavities in microwave domain that have very high Q values[6] and provide good isolation of the system from the background. Realization of superconducting circuits allowed very efficient free-space optics, while discovery of Josephson junctions provided nonlinear elements, allowing creation of designed 'atoms' in such circuits. Indeed, today some of the cleanest experimental results demonstrating entanglement in the electromagnetic field domain are done precisely with superconducting circuits. And some of the most advanced quantum computing platforms, including both quantum computers

---

[5] Quantum efficiency is

$$QE \equiv \frac{\text{\# of detected photons}}{\text{\# of incoming photons}}$$

[6] Quality factor of the cavity

$$Q \equiv 2\pi \frac{\text{energy stored}}{\text{energy lost in one cycle}}$$

developed by IBM and Google, are based precisely on superconducting circuits. So it is worth keeping in mind that while we talk about quantum optics, experiments are not done only in the visible range, but also in the microwave regime. Theoretical treatmant is the same for all electromagnetic fields, and might show difference only regarding to details of structure of 'atoms', either real or engineered as circuits, quantum dots, nitrogen-vacancy centers etc.

References

[1] Herschel W 1800 Experiments on the Solar, and on the Terrestrial Rays that Occasion Heat; With a Comparative View of the Laws to Which Light and Heat, or Rather the Rays Which Occasion Them, are Subject, in Order to Determine Whether They are the Same, or Different. Part I *Phil. Trans. R. Soc. Lond.* **90** 293

[2] Young T 1802 The Bakerian Lecture. On the theory of light and colours *Phil. Trans. R. Soc.* **92** 12

[3] Becquerel E 1949 L'image photographique colorée du spectre solaire *Comptes Rendues* **26** 181

[4] Taylor G I 1909 Interference fringes with weak light *Proc. Camb. Phil. Soc.* **15** 114

[5] Delbruck M 1980 Was Bose-Einstein statistics arrived at by serendipity? *J. Chem. Edu.* **57** 467

[6] Dirac 1927 The quantum theory of the emission and absorption of radiation *Proc. R. Soc. Lond.* A*114 243*

[7] Göppert- Mayer M 1931 Über Elementarakte mit zwei Quantensprüngen *Ann. Phys. (Leipzig)* **9** 273

[8] Bohm D and Aharonov Y 1957 Discussion of experimental proof for the paradox of Einstein, Rosen, and Podolsky *Phys. Rev.* **108** 1070

[9] Hanbury Brown R and Twiss R Q 1954 A new type of interferometer for use in radio astronomy *Phil. Mag.* **45** 663

[10] Hanbury Brown R and Twiss R Q 1956 A test of a new type of stellar interferometer on Sirius *Nature* **178** 1046

[11] Hanbury Brown R and Twiss R Q 1956 Correlation between photons in two coherent beams of light *Nature* **177** 27

[12] Glauber R J 1963 The quantum theory of optical coherence *Phys. Rev.* **130** 2529

[13] Bell J S 1964 On the Einstein Podolsky Rosen Paradox *Physics* **1** 195

[14] Morgan B L and Mandel L 1966 Measurement of photon bunching in a thermal light beam *Phys. Rev. Lett.* **16** 1012

[15] Kocher C A and Commins E D 1967 Polarization correlation of photons emitted in an atomic cascade *Phys. Rev. Lett.* **18** 575

[16] Burnham D C and Weinberg D L 1970 Observation of simultaneity in parametric production of optical photon pairs *Phys. Rev. Lett.* **25** 84

[17] Freedman S J and Clauser J F 1972 Experimental test of local hidden-variable theories *Phys. Rev. Lett.* **28** 938

[18] Carmichael H J and Walls D F 1976 Proposal for the measurement of the resonant Stark effect by photon correlation techniques *J. Phys. B: Atom. Molec. Phys.* **9** L43

[19] Kimble H J, Dagenais M and Mandel L 1977 Photon antibunching in reosnance fluorescence *Phys. Rev. Lett.* **39** 691

[20] Aspect A, Grangier P and Roger G 1981 Experimental tests of realistic local theories via Bell's theorem *Phys. Rev. Lett.* **47** 460

[21] Stoler S 1970 Equivalence classes of minimum uncertainty packets *Phys. Rev.* D*1 3217*

[22] Caves C M 1981 Quantum-mechanical noise in an interferometer *Phys. Rev.* D*23 1693*

[23] Wooters W K and Zureck W H 1982 A single quantum cannot be cloned *Nature* **299** 802

[24] Dieks D 1982 Communication by EPS devices *Phys. Lett.* **92A** 271

[25] Strickland D and Mourou G 1985 Compression of amplified chirped optical pulses *Optics Commun.* **56** 219

[26] Hong C K, Ou Z Y and Mandel L 1987 Measurement of subpicosecond time intervals between two photons by interference *Phys. Rev. Lett.* **59** 2044

[27] Carmichael H J 1993 Quantum trajectory theory for cascaded open systems *Phys. Rev. Lett.* **70** 2273

[28] Gardiner C W 1993 Driving a quantum system with the output field from naother driven quantum system *Phys. Rev. Lett.* **70** 2269

[29] Shor P W 1996 Fault-tolerant quantum computation *Proc. of 37th Conf. on Foundations of Computer Science*

[30] Linden N and Popescu S 2001 Good dynamics versus bad kinematics: is entanglement needed for quantum computation? *Phys. Rev. Lett.* **87** 047901

[31] Jozsa R and Linden N 2003 On the role of entanglement in quantum-computational speed-up *Proc. Math. Phys. Eng. Sci.* **459** 2011

[32] Clauser J F 1974 Experimental distinction between the quantum and field-theoretic predictions for the photoelectric effect *Phys. Rev.* D*9 853*

[33] Scully M O and Sargent M 1972 The concept of photon *Phys. Today* **25** 38

[34] Bloch F 1946 Nuclear induction *Phys. Rev.* **70** 460

[35] Feynman R P, Vernon F L and Hellwarth R W 1956 Geometrical representation of Schrödinger Equation for solving maser problems *J. Appl. Phys.* **28** 49

[36] Hahn E L 1950 Spin echoes *Phys. Rev.* **80** 580

[37] Purcell E M 1946 Spontaneous emission probabilities at radio frequencies *Phys. Rev.* **69** 681

[38] Lazio J and Deutsch L 2014 The deep space network at 50 *Phys. Today* **67** 31

[39] Niels Bohr Library & Archives, American Institute of Physics, College Park, MD USA, www.aip.org/history-programs/niels-bohr-library/oral-histories/31413-3

IOP Publishing

An Interactive Guide to Quantum Optics

Nikola Šibalić and C Stuart Adams

Chapter 3

What is quantum in quantum optics?

We have sought for firm ground and found none. The deeper we penetrate,
the more restless becomes the Universe; all is rushing about and vibrating
in a wild dance.

Max Born (1882–1970), Physics in my generation
(ed. Springer Verlag, 1969)

The essence of quantum optics involves quanta and interference—wave and particle. But there is some subtlety to the boundary between optics and quantum optics. For example, performing an optics experiment one photon at a time is not really quantum optics. So do we really need more than one quanta? Not necessarily, we can argue that making a single photon source and detecting that we do indeed have single photons is quantum optics. However, as we shall see single photon verification does require interference so our original state that we need both quanta and interference remains true.

To demonstrate quantumness we need quanta, and in quantum optics that means photons. So our starting point has to be to ask: What is a photon, how can we produce them, and how can we see them? We shall begin at the beginning with Einstein.

3.1 Einstein's introduction of the photon concept

Although the first hints of quantum physics appear in Max Planck's description of black-body radiation, it was Albert Einstein who in his 1905. paper 'Concerning an heuristic point of view toward emission and transformation of light' [1] established idea that light might be discrete, and argued for physical meaning of quanta based on experimentally verifiable consequences.

doi:10.1088/978-0-7503-2628-5ch3

Einstein starts with clearly delineating what was known in optics at the time, focusing on an area that might still exhibit new dynamics. He highlights that while wave theory was successful in describing diffraction, reflection, refraction, dispersion, these have been mostly observations of time averages rather than instantaneous values. But when it comes to emission and transformation of light, such as black-body radiation, fluorescence, ionization, these phenomena seemed to be more readily explained if one assumes that 'the energy of a light ray spreading from a point source is not continuously distributed over an increasing space but consists of a finite number of energy quanta which are localized at points in space, which move without dividing, and which can only be produced and absorbed as complete units.'

To arrive at this conclusion, Einstein relied on statistical physics and thermodynamic arguments, aiming to present argument that will be largely independent of underlying microscopic mechanisms that were unknown or at least uncertain at the time. Einstein sets out to calculate entropy S of the radiation in thermal equilibrium. The idea is that by comparing that expression with the expression of entropy of an ideal gas, one can draw a parallel and suggest that radiation is also behaving in a way like a gas. Suppose that we have radiation in thermal equilibrium[1] occupying a box of volume V limited by ideally reflective walls. We assume that all observable properties of such radiation can be then derived from knowledge of radiation density[2] $\rho(\nu)$ for a radiation of frequency ν. For radiation at thermal equilibrium, there is no transfer of heat between different frequency components, so each frequency can be considered to contribute independently to the total entropy

$$S = V \int_0^{+\infty} \varphi(\rho, \nu) \; d\nu,$$

where ϕ is some unknown function we will try to obtain. In thermal equilibrium, as reached by black-body radiation, entropy is maximized for a given total energy, that is, varying entropy around maximum gives to first order no change $\delta \int_0^{+\infty} \varphi(\rho, \nu) \; d\nu = 0$ provided we keep total energy change zero $\delta \int_0^{+\infty} \rho \; d\nu = 0$. From this it follows that $\int_0^{+\infty} \left(\frac{\partial \varphi}{\partial \rho} - C \right) \delta \rho \; d\nu = 0$ where C is independent of ν. Thus, $\frac{\partial \varphi}{\partial \rho}$ is independent of ν.

Using this independence, we can calculate then that adding energy dE increases the entropy of the system by

---

[1] To achieve thermal equilibrium, there has to be something in the box that would provide coupling between all the possible field's spatial modes and frequencies. These can be some atoms and molecules. Provided that there is such coupling, after some long time scale (determined by the microscopic details of the coupling in question) thermalization is achieved, i.e. there is no net (on average) transfer of energy between different spatial and frequency modes, although fluctuations persists as this is dynamical equilibrium on a micro level. If we are interested only in these average quantities, we don't need to consider underlying mechanisms that provide such coupling once the equilibrium is achieved.

[2] $\rho(\nu)d\nu$ is energy per unit volume of radiation in the frequency range between ν and $\nu + d\nu$.

$$dS = \int_{\nu=0}^{\nu+\infty} \left(\frac{\partial \varphi}{\partial \rho} \right) d\rho \ d\nu = \left(\frac{\partial \varphi}{\partial \rho} \right) dE.$$

At the same time, from the second law of thermodynamics we have $dS = \frac{1}{T} dE$, from which we have relation

$$\frac{\partial \varphi}{\partial \rho} = \frac{1}{T}. \tag{3.1}$$

From this one can calculate entropy, knowing that for $\rho \to 0$ entropy vanishes $\varphi \to 0$.

Wien [14] had found an earlier empirical relation

$$\rho = \alpha \nu^3 \exp(-\beta \nu / T),$$

that was in good agreement with the experiment for the high-frequency part of the black-body spectra $\beta \nu / T > 1$. Expressing T from that equation and replacing in equation (3.1), one obtains after integration

$$\varphi(\rho, \nu) = -\frac{\rho}{\beta \nu} \left[\ln \left(\frac{\rho}{\alpha \nu^3} \right) - 1 \right]. \tag{3.2}$$

In other words, for energy E of radiation with frequencies in the range between ν and $\nu + d\nu$, enclosed in volume V, total entropy is

$$S = V\varphi(\rho, \nu) d\nu = -\frac{E}{\beta \nu} \left[\ln \left(\frac{E}{V\alpha \nu^3 d\nu} \right) \right].$$

Thus, if we change the volume of the box containing radiation to V_0, the difference in two entropies is

$$S - S_0 = \frac{E}{\beta \nu} \ln \left(\frac{V}{V_0} \right). \tag{3.3}$$

We can now compare this with the result for ideal gas. If we consider Boltzmann's statistical interpretation of entropy S, it is a function $f(W)$ of probability W of a given state. When entropy increases, the system evolves towards a state of higher probability. Consider two systems with entropies $S_1 = f(W_1)$ and $S_2 = f(W_2)$. If these two non-interactive systems we consider as a single system whose entropy is $S = f(W_1 \cdot W_2) = S_1 + S_2$ then we know that entropy must be a function of a form

$$f(W) = k_B \ln(W) + \text{const}, \tag{3.4}$$

where constant k_B is Boltzmann constant. For ideal gas of N non-interacting particles in volume V the probability for possible configuration will be proportional to the volume each particle can independently explore, thus $W \propto V^N$. Therefore, change in entropy for gas held in volume V and V_0 is

$$S - S_0 = k_{\mathrm{B}} N \ln\left(\frac{V}{V_0}\right) \tag{3.5}$$

Comparing change of entropy for the ideal gas equation (3.5) with change of entropy for the radiation equation (3.3), it seems that radiation behaves thermodynamically as if consisting of a number of independent energy quanta of magnitude $\beta \nu k_{\mathrm{B}}$. Taking into account that in modern notation $\beta = h/k_{\mathrm{b}}$, quanta of light energy is $h\nu$ and we nowadays call these light quanta photons.

Einstein then proceeds to explain photoluminescence, give predictions of electron energies in photoelectric effect, and gas ionization. In spite of the clear explanations that this light quanta concept offered, it took time for the physics community to accept that wave theory of light indeed is incomplete. Max Planck, proposing Einstein for Prussian Academy membership in 1913, wrote 'That he may have missed the target in his speculations, as for example, in his hypothesis of light quanta, cannot really be held against him'. And even when Robert Millikan in 1916 experimentally confirmed Einstein's predictions regarding photoelectric effect, in the same paper he again voiced concern with its underlying quanta hypothesis saying 'I shall not attempt to present the basis for such an assumption, for, as a matter of fact, it has almost none at the time'. Since then, Einstein's hope that 'this approach may be useful to some investigators in their research' he expressed in his 1905. paper indeed paid off countless times.

3.2 New types of dynamics brought by quantum theory

Quantum theory did not solely add wave behaviour to the electrons whose dynamics was described before in analogy to point particles. Nor did it purely extend particle nature to light previously understood in analogy to waves. While quantum theory did introduced these 'wave mechanics' modifications, they could still be imagined in analogy to known systems. There were, however, three new sources that were introducing radically different dynamics, requiring new mathematical formalism, with no analogy in previously known physics. It is thanks to them that in the quantum world we can indeed do more than in the classical world known to physics before. These three sources are:

    (i) fundamentally different importance of measurement;
    (ii) state superpositions (particularly non-local or entangled states);
    (iii) identical nature of particles (particularly bosonic character of photons).

Measurements (i) we will discuss in detail in chapter 5. We note here that in contrast to classical mechanics, measurement of the system always modifies system dynamics. At the same time, all dynamics, that happens between the observations, is always continual. As we will see in chapter 5, even when a system can be detected only in one of two states, it always evolves *continuously* through superposition of states. Thanks to this continuous evolution, we can use interaction protocols that cancel information leakage about the system in unwanted degrees of freedom in environment, preserving coherence, as will be discussed in chapter 6. This continuous evolution of even two-state systems also allows states to continuously 'flow', tracking the current system eigenstates, as we modify external parameters, allowing adiabatic control. Finally, measurement has such strong feedback on the final projected states that it turns out to allow effective interactions in the post-selected states, implementing effective measurement-based interactions between photons, and allowing implementation of universal quantum gates using just linear optics elements and photon detectors [2]. These will be discussed in chapter 7.

Superpositions of a system in several several states (ii), whose total number scales exponentially with the number of the accessible states, are the key resource exploited in quantum algorithms to obtain a speed-up compared to classical algorithms [3, 4]. Non-local superpositions of two or more subsystems (e.g. spin of two electrons, or polarizations of the two photons) showed clearly that these entangled quantum systems cannot be reduced to the crowded carriage of classical physics that will reproduce the quantum predictions. Einstein, Rosen and Podolsky clearly focused attention of the community on this feature that becomes evident when one unites the new importance of measurements (i) with state superpositions (ii). The work of John Bell formalized their insight as a set of bounds on classically possible maximal correlations on measurements of two entangled subsystems. These Bell's inequalities have been violated in a number of experiments excluding the possibility that the world can be both *real* (outcome of the physical property of the system is determined by the physical properties of that system prior to and independent of measurement) and *local* (outcome of the physical property of the system can depend only on actions within the space-time cone allowed by relativity) at the same time. These will be discussed in chapter 8. Entanglement is exploited nowadays in protocols for quantum communications: quantum key distribution, quantum teleportation, quantum dense coding and more.

The identical nature of particles in quantum mechanics has important consequences on photons. Indistinguishability is important in calculating their statistical properties. In addition, as bosons, photons exhibit a tendency to stick together, as evident by the action of photon creation operator $\hat{a}^\dagger|n\rangle = \sqrt{n+1}\,|n+1\rangle$ for addition of a photon in the mode in state $|n\rangle$ that already contains n photons. Plank's black-body spectrum can be derived from this observation[3] and Einstein's A and B coefficients can be understood. Crucially this effective photon attraction also gives rise to the *nonlinearity* that is exploited in strongly-coupled atom–cavity systems that we will be looking at in chapter 9.

3.3 When are phenomena or technology quantum?

Quantum is an ever more popular adjective added to an array of technologies, products and phenomena. It is true that for the explanation of many common properties we need quantum theory. We have seen in chapter 1 that even under-standing sunshine in all it's spectral glory requires quantum theory. And many

---

[3] Based on this result Einstein derived the later statistic of ideal quantum gas, predicting Bose–Einstein condensate phase [5].

common technologies, like solid state lasers, modern hard disk drives, and MRI scanners in hospitals, all rely on aspects of quantum theory for their operation. Does that mean that everything we built using hard disk drives, or relying on sunshine, should be labelled as quantum? The adjective quantum would then lose all its informational value, since then everything is quantum in a way, constructed ultimately with (quantum) atoms and photons.

One can use quantum in a more restricted—and hence more informational—sense, as a label for particular phenomena, or a particular layer in a technology stack, that relies on new dynamics caused by the three sources (i)–(iii) described in section 3.2. More formally [6], if there exists some single-particle basis where the density operator describing relevant degrees of freedom is diagonal at all times, then phenomena or a new segment of technology stack are *classical*. This is because in such a single-particle basis dephasing has no effect on dynamics under consideration. Only when such a basis cannot be found do we rely on (ii) superpositions and/or (iii) quantum fluctuations until the (i) measurement event, and we can say that *particular* phenomena or technology layer are really using new features of quantum dynamics and hence are quantum, and not purely consisting of building blocks that are always quantum.

3.4 Single quanta, optics and biological systems

As a side note to the story of our human-centric history of quantum, it is worth remembering that even back in 1952, when Schrödinger (chapter 1) was expressing his disbelief that we will ever conduct experiments on single quanta and single molecules, our every cell in the organism was doing exactly this: each cell relies on a *single* DNA molecule as a recipe to cook up all the structural and functional building blocks. A few years earlier, in 1943. in Dublin, none other than Schrödinger delivered a lecture series entitled 'What is life?' [7] discussing physical principles behind biological processes, motivating the development of biophysics. Natural selection working over thousands and millions of years, often arrived at solutions that work close to bounds set by the laws of physics, as seen for example with the size of the composite eyes in insects whose size is limited by light diffraction[4]. How close to the quantum optics limits does Nature operate? There is experimental evidence that receptors in human eyes—although their detection efficiency is low, with peak quantum efficiency of $\approx 10\%$—can upon absorption of the single photon produce spikes in potential that will be perceived[5]. While perception of single quanta of light is impressive, this example does not rely on coherence, so it is not strictly a quantum phenomenon as defined in the preceding section 3.3. It was widely assumed that a hot and wet thermal environment of cells would not support coherent phenomena

---

[4] See for example figure 3 in [8]. For many other examples of biological systems and physical limits they often operate close to, see [9].

[5] Experimental evidence that the human eye can perceive single photons [13].

since any coherence would be quickly dephased. It turned out, however, that balance between coherence and the right amount of dephasing is achieved in plant photosynthesis complexes. In a classical random walk, traversed distance $\delta\mathbf{r}$ over time t evolves as $\langle(\delta\mathbf{r})^2\rangle \propto t$, where angular brackets in this case indicate ensemble average. On the other hand, a quantum random walk, when excitation is coherently coupled to identical systems to which it can hop proceeds faster, $\langle(\delta\mathbf{r})^2\rangle \propto t^2$. However, fully coherent transport is disadvantageous if the elements of the systems are heterogeneous, e.g. if every two-level element has slightly different detuning. In this case we can even have no transport at all, due to the phenomenon of **Anderson localization**. That happens because when calculating the probability of transfer between any two points, there are many paths through the neighbors, and if disorder is large, their contributions will be completely uncorrelated, and average to zero. This is a general wave phenomenon, but in the case of quantum particle random walk, destructive interference is the same.

In photosythesis complexes, photons are captured within the network of photoactive states. From there, energy excitation is coherently transferred between sites until it reaches the reaction center that converts energy excitation into chemical bond energy of sugars for storage. Immersed in water and among other molecules, the network of sites represents a heterogeneous environment of sites, and under fully coherent transport we would expect transport range to be very limited and slow. It turns out, however, that between a fully static heterogeneous environment and a rapidly fluctuating environment that would lead to classical random walk dynamics, there exists a region of thermal fluctuations that would keep dynamics coherent on short scales, but at the same time completely decohere long-range contributions, hence limiting Anderson localization type of destructive interference that prevents transport. In this intermediate regime transport is ultimately faster than what would be expected from fully classical random walk dynamics, keeping as it turns out a signature of underlying quantum dynamics even at room temperature [10]. In these disordered systems neither fully coherent nor fully dephased (classical) transport would achieve optimal speed of transport to the reaction centre.

Finally, it is worth reflecting on the fact that we humans are rather limited in seeing the world: only three different colors (at best), covering a small amount of the spectrum. All the figures and displays we can use in this book are limited to this small number of parallel information channels, as well as to our visual cortex that evolved to deal with these colors. It is very interesting that there exist animals that have much more interesting vision than us: mantis shrimps have as many as 12 different colour receptors [11], covering more of the spectrum, from UV to near infrared (300–720 nm). And if that were not enough, they can see circularly polarized light, and even actively twist their eyes (in addition to human-eye rotation up-down, left-right) to enhance contrast looking at polarized objects [12]. Mantis shrimps probably evolved this vision to better see transparent sea living organisms that they prey on. These might be transparent, but they are full of chiral sugars that are clearly visible to mantis shrimps. One can only wonder whether if humans had such eyesight, would we have discovered quantum optics even earlier, and how many communications channels we could have played with for our figures and displays.

References

[1] Arons A B and Peppard M B 1965 Einstein's proposal of the photon concept–a translation of the Annalen der Physik paper of 1905 *Am. J. Phys.* **33** 367

[2] Knill E, La R and Milburn G J 2001 A scheme for efficient quantum computation with linear optics *Nature* **409** 46

[3] Shor P W 1997 Polynomial-time algorithms for prime factorization and discrete logarithms on a quantum computer *SIAM J. Comput.* **26** 1484

[4] Grover L K 1997 Quantum mechanics helps in searching for a needle in a haystack *Phys. Rev. Lett.* **79** 325

[5] Bose S 1974 Planck's law and the light quantum hypothesis [translation into English of 1924 paper] *Am. J. Phys.* **44** 1056

[6] Johnson T H, Clark S R and Jaksch D 2014 What is a quantum simulator? *EPJ Quantum Technol.* **1** 1

[7] Schrödinger E 1944 *What Is Life? The Physical Aspect of the Living Cell* (Cambridge: Cambridge University Press)

[8] Barlow H B 1952 The size of ommatidia in apposition eyes *J. Exp. Biol.* **29** 667

[9] Bialek W 2012 *Biophysics: Searching for Principles* (Princeton, NJ: Princeton University Press)

[10] Mohseni M, Rebentrost P, Lloyd S and Aspuru-Guzik A 2008 Environment-assisted quantum walks in photosynthetic energy transfer *J. Chem. Phys.* **129** 174106

[11] Thoen H H, How M J, Chiou T-H and Marchall J 2014 A different form of color vision in mantis shrimp *Science* **343** 411

[12] Daly I M, How M J, Partridge J C, Temple S E, Justin Marshall N, Cronin T W and Roberts N W 2016 Dynamic polarization vision in mantis shrimps *Nat. Commun.* **7** 12140

[13] Tinsley J N, Molodtsov M I, Prevedel R, Wartmann D, Espigulé-Pons J, Lauwers M and Vaziri A 2016 Direct detection of a single photon by humans *Nat. Commun.* **7** 12172

[14] Wien W 1897 XXX. On the division of energy in the emission-spectrum of a black body *Lond. Edin. Dub. Phil. Mag. J. Sci.* **43** 214–20

Part I

One quanta

IOP Publishing

An Interactive Guide to Quantum Optics

Nikola Šibalić and C Stuart Adams

Chapter 4

One quanta

4.1 Introduction

Maxwell's introduction of the field concept. Traditionally, quantum optics textbooks begin with the 'classical' electromagnetic (EM) fields, then introduce the idea that light is made of photons, and so we need to quantize the field. But this distinction between classical and quantum can be somewhat misleading—especially as it is tempting to think of classical as normal and quantum as weird. Historically, Maxwell's introduction of the field concept was, in fact, a dramatic leap away from our 'classical' worldview. As Einstein wrote [1],

> *Before Maxwell people thought of physical reality—in so far as it represented events in nature—as material points, whose changes consist only in motions which are subject to total differential equations. After Maxwell they thought of physical reality as represented by continuous fields, not mechanically explicable, which are subject to partial differential equations. This change in the conception of reality is the most profound and the most fruitful that physics has experienced since Newton.*

 LabBricks literature graph for this chapter can be found at https://labbricks.com/#10/q3zxIvyL1v

doi:10.1088/978-0-7503-2628-5ch4

Maxwell's EM field consists of modes: photons are simply the number of energy quanta in each mode. In quantum optics, the solutions of Maxwell's field equations form the **modes**. Photons are simply quanta of energy that populate these modes. The modes are key, whether we populate them less so. For example, in quantum optics, even the vacuum mode is important. It is the modes that contain all the key information about phase and coherence. In contrast, the photons tell us not much, only how many clicks we might expect on our photon counters. Consequently, when we talk about a photon we should also talk about the mode that this photon occupies. Here, in Part I we shall begin with one quanta in one mode, then in Part II we introduce a second—two quanta—and finally discuss many.

Photons can be thought of as prototypical two-state quantum systems or qubits. Part I is organized as follows: first, in section 4.2 we address the question, *What is a photon?* Mathematically, many of the interesting properties of single photons, in particular, the mathematics of the single-photon emitter, and the theory of how to describe single-photon interference can be treated using a *two-state quantum mechanics*. For the emitter, the two states are two energy levels. For interference, the two states are two paths. Although the two cases are very difficult, the mathematics is identical. Consequently, we shall spend some time developing the quantum mechanics of two-state systems. These ideas also provide the foundations of **quantum computing** so are increasingly applicable and useful. To visualize two-state quantum dynamics we shall introduce the **Bloch and Poincaré spheres**, and **density matrices**. We shall extend the two-state model to include interaction with the environment and introduce the concept of decoherence. We also mention measurement tools such as homodyne detection, and visualization tools such as the Wigner function. When we need to consider more than one quanta in Part II we shall also introduce additional theoretical tools such as the Jaynes–Cummings model.

4.2 What is a photon?

For over one hundred years physicists have struggled with the question, *What is a photon?* It is often reported that in 1951 Einstein said, *All the fifty years of conscious brooding have brought me no closer to the answer the question, 'What are light quanta?' Of course today every rascal thinks he knows the answer, but he is deluding himself.* Individual photons, in their ideal form, do not exist in Nature. Physicists have to employ advanced technologies such as lasers, non-linear optics, or exotic quantum materials to produce them. Although aspects of single-photon emission and measurement continue to challenge our reasoning[1], experimental quantum optics provides ample and convincing evidence for their existence. As experiments on individual photons became accessible, physicists began to reference their existence to detection events. For example, Roy Jay Glauber (New York City 1925–Newton 2018)—co-recipient of the 2005 Nobel Prize in Physics for his contribution to the quantum theory of optical coherence—said [2]

A photon is what a photon detector detects.

---

[1] Feynman on where do photons come from http://www.youtube.com/watch?v=eebWoZkN3FQ.

and

> *A photon is where a photon detector detects it.*

Similarly, Feynman writes in *QED: The strange theory of light and matter* [3],

> *We know light is made of particles because we can take a very sensitive instrument that makes clicks when light shines on it, and if the light gets dimmer, the clicks remain just as loud—there are just fewer of them.*

But clicks are not a sufficient proof of the reality of photons. As we shall see, evidence for the quantization of the EM field and hence the existence of photons comes from experiments that generally involve both counting and interference. Perhaps the clearest demonstration of the special character of photons is the measurement of the **single-photon Wigner function using homodyne detection** [4, 5], to be discussed in chapter 5 on measurement.

4.3 Photon properties

What are key properties of a photon? A photon is a single excitation (or quanta), with an energy $h\nu$, where ν is the centre frequency and h is Planck's constant. But an excitation of what? It is a single excitation of a particular EM field mode, where a mode is a solution to Maxwell's field equations. For this reason we might assign most of the properties of the photon as properties of the mode. In general, a photon in a different mode is a different photon, however, if two photons are indistinguishable apart from their propagation direction we may be able to combine them on a beam splitter. In space, the mode defines the effective shape of the photon—how long it is and how wide it is. For example, the spatial mode could be a Gaussian beam like a laser beam, or it could be a dipolar radiation pattern. On top of the spatial distribution, we also have some time dependence which is related to spectral properties of the photon. In time, we might have a Gaussian pulse or an exponentially decaying pulse. The time evolution of the field might be a coherent sum of frequencies or a statistical mixture (incoherent sum) of frequency components. The mode also includes information about the polarization of the EM field[2]. In general photons are complicated because they can have an infinite number of degrees of freedoms. In quantum optics, physicists try to work with simple cases such as photons in cavities which have a single well-defined spatial mode.

Photon number (Fock) states. In theoretical quantum optics, we begin with modes, and then create photons within them. All photons in the same mode are by definition identical and indistinguishable. To represent our mode we use a state vector, $|n\rangle_{\text{mode}}$, where n is the number of photons in the mode. The basis where we refer to the number of photons in a particular mode is known as the **Fock basis**.

---

[2] Photons may also have more degrees of freedom like orbital angular momentum, see e.g. [6] but we shall not discuss that here.

Photon number (Fock) states have similar properties to the states of the quantum harmonic oscillator. A big difference between classical and quantum is that, whereas in classical physics a quantity like an electric field is well-defined, in quantum physics it is not. When we measure a quantum field, such as the electric field of light, we obtain a distribution of values, see section 4.16. The variance of this distribution is often referred to as **quantum noise**, but it is not like classical noise, it is more of an indeterminacy like we encounter in Heisenberg uncertainty relations. For light, the distribution of electric field values is given by the same formulae as the distribution of particle positions for a harmonic oscillator[3]. This follows from the similarly of the wave equation for light, $\ddot{\mathcal{E}} + \omega^2 \mathcal{E} = 0$, where $\mathcal{E}$ is the electric field, and the equation of a harmonic oscillator, $\ddot{x} + (\kappa/m)x = 0$, where x is the particle displacement.

Photon creation in theory. When we write down a Fock state such as $|n\rangle_{\text{mode}}$, we are neglecting all the information about the mode. A mode with no photons is called the **vacuum state**. We write this as $|0\rangle_{\text{mode}}$. To add a photon in this mode, we use a **creation operator**, $\hat{a}^\dagger$. The creation and annihilation operators, $\hat{a}^\dagger$ and $\hat{a}$, respectively, are used to describe the addition or subtraction of one quanta of energy. They are often introduced in the context of the simple harmonic oscillator which has the required feature that the spacing between level with n and $n+1$ quanta is independent of n. Some useful properties of these operators include:

$$\hat{a}^\dagger|n\rangle = \sqrt{n+1}\,|n+1\rangle,$$
$$\hat{a}|n\rangle = \sqrt{n}\,|n-1\rangle,$$
$$\hat{a}^\dagger \hat{a}|n\rangle = n|n\rangle.$$

A single-photon state is generated by applying the creation operator to the vacuum state, i.e.

$$|1\rangle_{\text{mode}} = \hat{a}^\dagger |0\rangle_{\text{mode}}.$$

Mostly, we shall drop the subscripts, but sometimes we will need them because we want to describe photons in different modes. Or alternatively, instead of subscripts we could use distinct creation operators, $\hat{a}^\dagger$, and $\hat{b}^\dagger$, to create photons in different modes. Before considering how to create photons using a single-photon emitter, see section 4.12, we shall discuss how we can think of photons as qubits, and borrow from the productive seam of concepts that are used in quantum information processing (QIP) and quantum computing.

---

[3] For a Fock state with n photons in a mode with angular frequency ω and volume V, the probability of measuring an electric field, $\tilde{\mathcal{E}} = \mathcal{E}/\sqrt{\hbar\omega/\epsilon_0 V}$ is

$$P(\tilde{\mathcal{E}}) = \frac{1}{\sqrt{\pi}\,2^n n!}|H_n(\tilde{\mathcal{E}})|^2 e^{-\tilde{\mathcal{E}}^2}.$$

4.4 Photons as qubits

A **qubit** is any two-state quantum system. Two-state quantum mechanics is the starting point for many of the concepts in quantum optics. A two-state model applies to: the energy levels involved in single-photon emission, the state of polarization of a photon (or any coherent light field), photon transmission or reflection at a beam splitter. Any two-state system may be considered as a candidate for the *qubits* in a quantum computer. Examples including photons, atoms, ions, molecules, or artificial atoms (two-level systems such as spin-1/2 impurities in semiconductors or resonant LC circuits in superconductors). For photonic quantum information, the two states may be two polarizations. This is known as **polarization encoding**. Correlation between the polarizations of a photon pair is also the basis of quantum communication schemes like **quantum cryptography**. Alternatively, the two states could be two spatial modes. This is known as **dual-rail encoding**. We shall discuss both in this section.

As the field of quantum information has evolved, it has become increasingly fruitful to treat the photons in quantum optics as a type of qubit. In quantum information theory, the two states are typically labelled $|0\rangle$ and $|1\rangle$. The labels $|0\rangle$ and $|1\rangle$ typically represent two energy levels such a ground and excited state,

$$|0\rangle \equiv |g\rangle \ \text{ and } \ |1\rangle \equiv |e\rangle.$$

But in quantum optics, there are no energy levels as such, and $|0\rangle$ and $|1\rangle$ are used to represent something different. As we mentioned for photons, a qubit may be encoded in two different ways.

Polarization qubits. As a photon has two states of polarization, photons are **polarization qubits**. The two orthogonal polarizations could be left- and right-circular, and our qubit states are

$$|0\rangle \equiv |\circlearrowleft\rangle \ \text{ and } \ |1\rangle \equiv |\circlearrowright\rangle.$$

Or we could use orthogonal linear polarizations, e.g. $|0\rangle \equiv |H\rangle$ and $|1\rangle \equiv |V\rangle$, where H and V denote horizontal and vertical, respectively. Exploiting the polarization degree of freedom in this way, is known as **polarization encoding**.

Spatial mode (dual-rail) photonic qubits. Alternatively, we can also use modes as basis states. For example, a photon in spatial mode '0' or '1' can be used to represent state $|0\rangle$ and $|1\rangle$. This is known as **dual-rail encoding**. The two modes need to be distinguishable. Typically, we might use two spatially separate paths, or two wave guide tracks. A **polarizing beam splitter (PBS)** converts between polarization encoding and dual-rail encoding. In figure 4.1 we show how a PBS cube splits the horizontal and vertical components into two spatial modes.

The state vector. In any encoding scheme, we represent the qubit using a state vector,

$$|\psi\rangle = a|0\rangle + b|1\rangle, \tag{4.1}$$

where a and b are complex coefficients that may be time dependent. According to the Born rule $|a|^2$ and $|b|^2$ are the probabilities to be detected in $|0\rangle$ and $|1\rangle$, respectively.

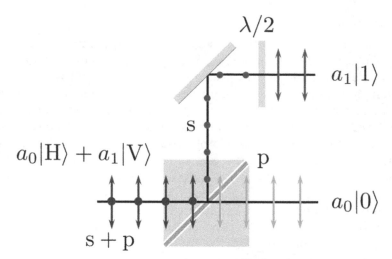

Figure 4.1. Mapping between polarization-encoded photons with state vector, $|\psi\rangle = a_0|\mathsf{H}\rangle + a_1|\mathsf{V}\rangle$, and a dual-rail encoded photon with state vector, $|\psi\rangle = a_0|0\rangle + a_1|1\rangle$. The s and p labels refer to polarization perpendicular (from the German 'senkrecht') and parallel to the plane of propagation. The $\lambda/2$ wave plate is aligned at 45° to induce a half-wave ($\lambda/2$) relative phase shift between diagonal and anti-diagonal components. This rotates the plane of polarization by 90°. In photonic quantum computing, this converts a polarization qubit into a dual-rail qubit.

If our quantum system has to be in one of only two states then $|a|^2 + |b|^2 = 1$ and the 'length' (in Hilbert space) of the state vector is unity $\langle\psi|\psi\rangle = 1$. In *quantum computing*, equation (4.1) is known as the *qubit* state vector. Note that a and b are continuous 'analogue' variables and can be specified with infinite precision. Implicit in the **pure state** description is both that we have infinite knowledge, but still do not know everything. The coefficients a and b may be specified to infinite precision, but if they are both non-zero then still we do not know, and cannot know, whether the system will be detected in state $|0\rangle$ or $|1\rangle$. This non-reality of the mathematical description used in quantum mechanics is succinctly communicated in this Phillip Ball quote,

> *The wave function [or state vector] is NOT a description of the quantum object, it is a prescription for what to expect when we make a measurement on that object.* Phillip Ball, *Quantum mechanics is not weird*, Qiskit Seminar, 12th February 2021.

The mathematical idealization of a pure state is like a plane wave in optics, mathematically convenient but only an approximate representation of reality. How close we can get in reality to preparing any desired pure state is referred to as the **fidelity**. As experimental techniques improve, we move asymptotically closer towards the production of pure states in the lab. For demanding applications such as quantum computing, state purity of over 99% is desirable. But even if we are able to achieve a high fidelity, we may still be interested in cases where things do not follow the pure state description. To go beyond pure states, it is convenient to use a **density matrix** description first introduced by Johann von Neumann in 1927 [7]. As we shall see, in section 4.8 and chapter 6 in particular, the density matrix description is particularly convenient when we need to treat the interaction between our quantum system and the environment, and how this leads to the phenomenom of decoherence. Next, before we can say more about photons as polarization-encoded qubits, we should talk a bit more about polarization.

4.5 Photon polarization

Figures 4.2 and 4.3 show a schematic of left- and right-circularly polarized photons, respectively. There is no universally agreed convention as to what should call left and right. We define left-circular, as an electric field vector that traces out a left-handed helix in space, i.e. a left-handed screw, see figure 4.2. The handedness is referenced to the direction of propagation. In this case, a left-circular photon has an angular momentum $+\hbar$ in the propagation direction. If we look into the beam we

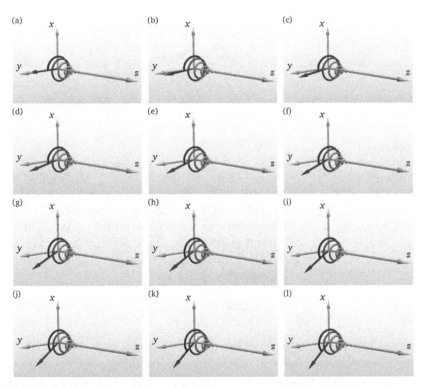

Figure 4.2. Left-circularly polarized photon. In this labelling convention, the tip of the electric field vector traces out a left-handed screw in space, respectively. Using your left hand with your thumb pointing in the propagation direction, your fingers point in the direction of the twist in the electric field. Panels (a)–(l) show the photon moving in time. Looking into the field (observer perspective) the electric field vector in the $z = 0$ plane rotates anti-clockwise in time. An interactive figure is available at http://doi.org/10.1088/978-0-7503-2628-5.

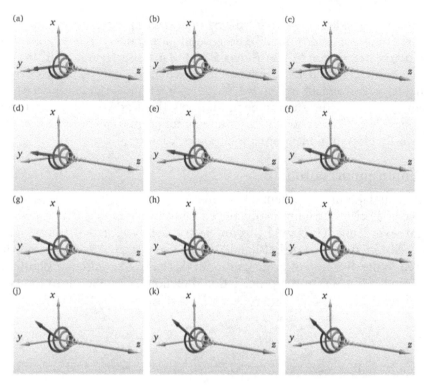

Figure 4.3. Right-circularly polarized photon. The tip of the electric field vector traces out a right-handed screw in space, respectively. (a)–(l) The electric field vector in the $z = 0$ plane rotates clockwise in time. An interactive figure is available at http://doi.org/10.1088/978-0-7503-2628-5.

observe the electric field vector rotating anti-clockwise, see [8] for more details. Figure 4.4 shows the time evolution of the basis states (columns 1 and 2) and an equal superposition of both (column 3). The frames show the field vector in the xy plane at $z = 0$ for light propagating in the $+z$ direction. From the observer perspective, figure 4.4(a), i.e. sitting at positive z looking back towards the origin, the left-circular vector rotates anti-clockwise and the right-circular vector rotates clockwise. We can write the superposition state as

$$|\psi\rangle = \frac{1}{\sqrt{2}}(|\circlearrowleft\rangle + e^{i\phi}|\circlearrowright\rangle) \equiv \frac{1}{\sqrt{2}}(|0\rangle + e^{i\phi}|1\rangle).$$

where ϕ is the relative phase between the two components. For the relative phase, ϕ, we choose an example with $\phi = \pi/2$. This means that time evolution of the $|1\rangle$ state lags behind $|0\rangle$ by a quarter of a period. An equal superposition of left and right given a linear polarization along the $-45°$ diagonal in the xy plane. This illustrates how the circular and linear basis are related and hence interchangeable.

If instead we take the source perspective, sitting at negative z and looking in the $+z$ direction then the x flips and the rotation directions flip. So now the left- and right-circular vectors appear to rotate clockwise and anti-clockwise, respectively.

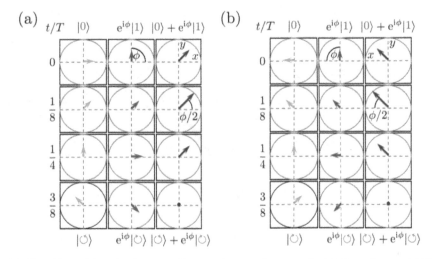

Figure 4.4. The effect of changing between observe and source frame on the labelling of states. (a) From the perspective of the observer, the electric field vector for left- and right-circular light, labelled $|\circlearrowleft\rangle$ and $|\circlearrowright\rangle$, rotate anti-clockwise and clockwise, respectively. (a) From the perspective of the source, i.e. looking in the positive z direction, the rotation directions are opposite.

Consequently, as a general rule we should always specify both the propagation direction plus whether we are using the source or observer perspective.

4.6 The Bloch and Poincaré sphere

To represent the state of polarization of a photon, or the state vector of a photonic qubit (for either polarization or dual-rail encoding) we use the **Poincaré sphere** and **Bloch sphere**, respectively. The Poincaré sphere—which pre-dates the Bloch sphere —was initially introduced to describe phase trajectories for a two-dimensional coupled system [9]. Subsequently, Poincaré applied the concept to the polarization of light [10]. As a photon has two polarization states, see figures 4.2 and 4.3, the mathematics of single-photon polarization is identical to other two-state systems (qubits). Hence, there is a one-to-one mapping between the Poincaré sphere used to represent photon polarization and the Bloch sphere used to represent the state of any two-state quantum system.

In the Bloch/Poincaré sphere representation, we write the state vector in the $\{|0\rangle, |1\rangle\}$ or $\{|\circlearrowleft\rangle, |\circlearrowright\rangle\}$ basis as

$$|\psi\rangle = \begin{pmatrix} \cos\dfrac{\theta}{2} \\ e^{i\phi}\sin\dfrac{\theta}{2} \end{pmatrix}, \tag{4.2}$$

where the angles θ and ϕ can be used to represent the state vector as a point on a unit sphere, see figure 4.5. The arrow pointing from the origin to the point (θ, ϕ) is

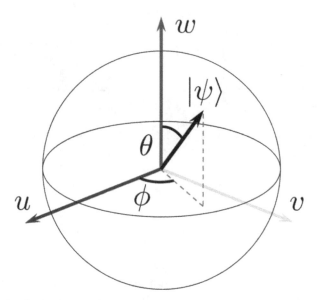

Figure 4.5. The Bloch or/and Poincaré sphere: the Bloch vector $|\psi\rangle$ is drawn from the origin to the point (θ, ϕ) on a unit sphere.

known as the *Bloch vector*. As we are dealing with a two-state system, the mathematics is the same as a spin $-1/2$ and we can relate the cartesian components of the Bloch vectors, labelled u, v, and w, to the expectation values of Pauli spin matrices, see section A.1 in the appendix. The w-component of the Bloch vector is equal to the 'population' difference between states $|0\rangle$ and $|1\rangle$, i.e. $|a|^2 - |b|^2$. Unit vectors in the $\pm w$ directions, North or South poles, correspond to all the populations in the $|0\rangle$ or $|1\rangle$ state, respectively. The u and v-components are known as the **coherences**, as they tell us about the relative phases of the two states. u and v correspond to the real and imaginary parts of the coherence, respectively[4]. Points on the equator, $\theta = \pi/2$, have a state vector

$$|\psi\rangle = \frac{1}{\sqrt{2}}(|0\rangle + e^{i\phi}|1\rangle),$$

corresponding to equal superpositions of $|0\rangle$ and $|1\rangle$. Such states have maximal coherence, with the relative phase between the two basis states given by the azithmuthal angle, ϕ. To represent incoherence we will need to go beyond the state vector description, see section 4.8.

---

[4] Note that as the Bloch vector pre-dates quantum computing some books use a different definition. In some cases, a factor of $\frac{1}{2}$ is absorbed into the definition of u and v, and the sphere is flipped such that $|1\rangle$ is at the North pole which changes the sign of w and v. However the definitions used here are standard in the quantum computing literature.

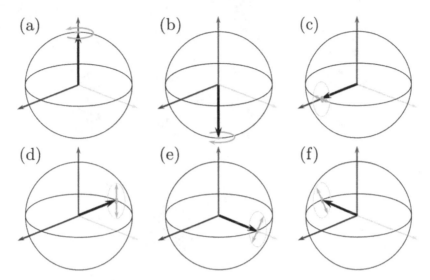

Figure 4.6. The Poincaré sphere where the $|0\rangle$ and $|1\rangle$ states (North and South pole) correspond to left- and right-circular polarizations, (a) $|0\rangle \equiv |\circlearrowleft\rangle$ and (b) $|1\rangle \equiv |\circlearrowright\rangle$, respectively.

If we chose the circular basis and associate $|0\rangle$ with left-circular polarization, we write the polarization state vector as

$$|\psi\rangle = \cos\frac{\theta}{2}|\circlearrowright\rangle + e^{i\phi}\sin\frac{\theta}{2}|\circlearrowleft\rangle,$$

where the blue and red arrows correspond to left- and right-circular, respectively (from the observer perspective). Linear polarization corresponds to an equal superposition of left and right-circular polarization, as illustrated in figure 4.4.

The six positions in the Bloch sphere shown in figures 4.6 and 4.7, correspond to the following polarizations: (a) $|0\rangle = |\circlearrowleft\rangle$, (b) $|1\rangle = |\circlearrowright\rangle$, (c) $|+\rangle_x = \frac{1}{\sqrt{2}}(|\circlearrowleft\rangle + |\circlearrowright\rangle)$, (d) $|-\rangle_x = \frac{1}{\sqrt{2}}(|\circlearrowleft\rangle - |\circlearrowright\rangle)$, (e) $|+\rangle_y = \frac{1}{\sqrt{2}}(|\circlearrowleft\rangle + i|\circlearrowright\rangle)$, (f) $|-\rangle_y = \frac{1}{\sqrt{2}}(|\circlearrowleft\rangle - i|\circlearrowright\rangle)$.

(a) The North pole

$$|0\rangle = |\circlearrowleft\rangle.$$

With the caveat, that whether we call this left- or right-circular depends on whether we take the observer or source perspective.

(b) The South pole

$$|1\rangle = |\circlearrowright\rangle.$$

Note that opposite points (antinodal points) anywhere on the Bloch sphere are orthogonal, i.e.

$$\langle 0|1\rangle = \langle\circlearrowleft|\circlearrowright\rangle = 0.$$

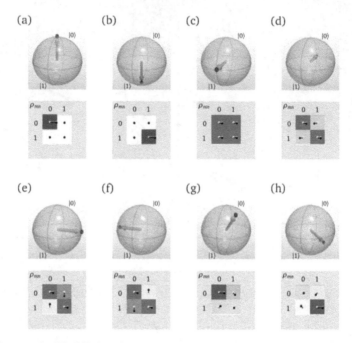

Figure 4.7. Points on the Bloch/Poincaré sphere and the corresponding density matrices: (a) $|0\rangle$, (b) $|1\rangle$, (c) $|+\rangle_x$, (d) $|-\rangle_x$, (e) $|+\rangle_y$, (f) $|-\rangle_y$. (g) and (h) show a general point $|\psi\rangle = \cos(\theta/2)|0\rangle + e^{i\phi}\sin(\theta/2)|1\rangle$ with $(\theta, \phi) = (\pi/4, \pi/4)$ and $(\theta, \phi) = (3\pi/4, 3\pi/4)$, respectively. An interactive figure is available at http://doi.org/10.1088/978-0-7503-2628-5.

(c) The equator with azithmual (or longitudinal) angle $\phi = 0$, i.e. along the $+x$ axis,

$$|+\rangle = \frac{1}{\sqrt{2}}(|\circlearrowleft\rangle + |\circlearrowright\rangle).$$

This corresponds to linear polarization and is often denoted as $|H\rangle$ for horizontal.

(d) The equator with azimuthal (or longitudinal) angle $\phi = \pi$, i.e. along the $-x$ axis,

$$|-\rangle = \frac{1}{\sqrt{2}}(|\circlearrowleft\rangle - |\circlearrowright\rangle).$$

This corresponds to a linear polarization orthogonal to $|+\rangle$ so if $|+\rangle = |H\rangle$ then we can label $|-\rangle$ as $|V\rangle$ for vertical.

(e) The equator with azimuthal (or longitudinal) angle $\phi = \pi/2$, i.e. along the $+y$ axis,

$$|+i\rangle = \frac{1}{\sqrt{2}}(|\circlearrowleft\rangle + i|\circlearrowright\rangle).$$

This corresponds to a linear polarization at 45° to the x and y axes. We shall call this $|\nearrow\rangle$. Whether the angle of the polarization axis is $\pm45°$ depends on our choice of sign convention for left- and right. Figure 4.4 illustrates how our choices are affected if we take a source or observer perspective.

(f) The equator with azithmual (or longitudinal) angle $\phi = -\pi/2$, i.e. along the $-y$ axis,

$$|-\mathrm{i}\rangle = \frac{1}{\sqrt{2}}(|\circlearrowright\rangle - \mathrm{i}|\circlearrowleft\rangle).$$

This corresponds to a linear polarization at 45° to the x and y axes. We shall call this $|\searrow\rangle$.

4.7 Photon polarization: linear basis

So far, we have described photon polarization using the circular basis, where $|0\rangle = |\circlearrowright\rangle$ and $|1\rangle = |\circlearrowleft\rangle$. In optics, it is common to work in a linear basis. Consequently, it may be convenient to rotate the Poincaré sphere such that $|H\rangle$ and $|V\rangle$ are at the North and South poles, respectively, as in figure 4.8, see, e.g. [11]. In this case, points on the equator correspond to a superposition of orthogonal linear polarizations

$$|\psi\rangle = \frac{1}{\sqrt{2}}(|H\rangle + \mathrm{e}^{\mathrm{i}\phi}|V\rangle).$$

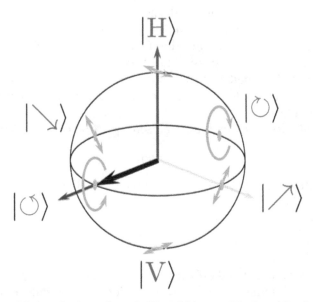

Figure 4.8. The rotated Poincaré sphere where the $|0\rangle$ and $|1\rangle$ states (North and South pole) correspond to horizontal, $|H\rangle$, and vertical, $|V\rangle$, linearly polarized light, respectively.

From figure 4.8 we see that changing the relative phase ϕ switches the polarization between from left-circular to diagonal to right-circular to anti-diagonal and back. In optics, this relative phase can be changed using a linearly birefringent medium such as quartz. As we have seen, polarization optics such as the PBS, figure 4.1, convert between the polarization basis $\{|\mathsf{H}\rangle, |\mathsf{V}\rangle\}$ and spatial modes, and subsequently, the vertical polarization is converted to horizontal using a half-wave ($\lambda/2$) plate. In the language of quantum computing, the PBS plus $\lambda/2$ converts a polarization-encoded qubit into a *dual-rail* encoded qubit. Whereas polarization encoding is convenient for quantum communication schemes like quantum cryptography, dual-rail encoding is more convenient for quantum computating, in particular using photonic chips, see e.g. [11]. We shall discuss this in more detail both later in this chapter and in part II.

4.8 Density matrix

Often states of light and matter do not have perfect coherence. Typically, we have only partially coherence which means that the phase of the field is not well defined. In this case the pure state description based on state vectors is insufficient. This partial coherence may arise when our quantum system is coupled to the environment. The theory of two-level systems coupled to the environment was originally developed to describe the dynamics of spins in nuclear magnetic resonance (NMR), see F Bloch Nobel lecture and E M Purcell Nobel lecture. But the same ideas can also be applied to a photon in a mixture of modes. In cases where we do not know everything about the state or the relative phase between basis states we use a **mixed state** description based on the **density matrix**. We mention it now, such that we can get used to visualizing the state of a qubit using both the Bloch/Poincaré sphere and the density matrix.

Another motivation for working with the density matrix is that is can be used to provide a complete desription of more than one two-state system (more than one qubit). This is important as we cannot simply use multiple Bloch spheres to represent multiple qubits because of the possibility of **entanglement**. We shall discuss this in more detail in Part II. Initially we shall introduce the density matrix along side the Bloch sphere for a pure state, and then go on to consider decoherence and mixed states in later chapters.

For a pure state described by the state vector $|\psi\rangle$, the density matrix is defined as

$$\rho = |\psi\rangle\langle\psi|. \tag{4.3}$$

This can be written in terms of the components of the Bloch vector as (see appendix A.2)

$$\rho = \frac{1}{2}\begin{pmatrix} 1 + w & u - iv \\ u + iv & 1 - w \end{pmatrix}. \tag{4.4}$$

The diagonal terms are real and only depend on the polar angle of the Bloch vector, θ. The off-diagonal terms have the same θ-dependence but rotate in opposite directions in the complex plane as a function of ϕ. Figure 4.7 shows the Bloch vector and corresponding density matrix for different points on the Bloch sphere. To visualize the density matrix we use a rainbow colour wheel in the complex plane.

Red is the positive real axis, yellow positive imaginary, green negative real and blue negative imagining. Also, we plot the coordinates of each matrix element in the complex plane—a phasor representation. In the interactive figures, the reader can decide whether they find the dots or colours more convenient. Below we show both the Bloch sphere and density matrix visualization. To illustrate decoherence and in Part II for two photons, where we can no longer use the Bloch sphere representation, the density matrix visualization becomes particularly useful.

4.9 A photon beam splitter

Now that we have introduced the state vector and density matrix for our photonic qubit we can talk about how we might introduce a change of state. Our first example is a beam splitter in a dual-rail encoding scheme. In figure 4.9 we show two species of beam splitter, both with two inputs and two outputs, i.e. dual rails[5]. First, in figure 4.9(a) we show a bulk optics schematic, where the reflective surface is a dielectric coating at the interface between two glass prisms. Second, in figure 4.9(b) we show a schematic of photonic wave guide beam splitter, see also [12]. In this section, we consider the bulk **beam splitter** shown in figure 4.9(a). We shall find that in the language of quantum information this corresponds to a rotation of the Bloch/Poincaré vector.

Using dual-rail encoding, we write the state vector of the photon as

$$|\psi\rangle = \begin{pmatrix} a_1^\dagger \\ a_2^\dagger \end{pmatrix},$$

where $a_j^\dagger$ is a creation operator that generates a photon in spatial mode j. The operator $a_j^\dagger$ creates a photon in mode j with probability $|a_j|^2$. Throughout this section we shall use the Fock basis—first introduced in section 4.3, where $|n\rangle_j$ represents n

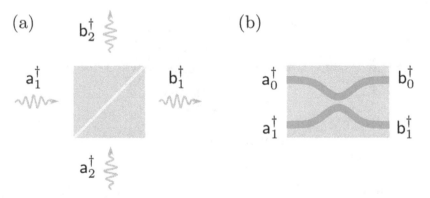

Figure 4.9. The input and output modes of a beam splitter. (a) Bulk optics: the operator $a_j^\dagger$ creates a photon in mode j probability $|a_j|^2$. Using the Fock basis where $|n\rangle_j$ represents n photons in mode j, $a_j^\dagger|0\rangle_j = a_j|1\rangle_j$. (b) Photonic wave guide beam splitter.

---

[5] For additional discussion on aspect of beam splitters, see e.g. [13].

photons in mode j, and the action of the creation operator on an empty mode is described by $a_j^\dagger |0\rangle_j = |1\rangle_j$.

Now consider what happens when the two modes are incident on a glass interface that is partially reflecting, see figure 4.9(a). The state vector after the beam splitter is written as

$$|\psi\rangle_{\text{out}} = \begin{pmatrix} b_1^\dagger \\ b_2^\dagger \end{pmatrix}. \tag{4.5}$$

Note that we are deliberately using distinct labels for the input and output modes. The $a_j^\dagger$ and $b_j^\dagger$ operators create a photon for the input and output channels, respectively. If there is no loss then $|\psi\rangle_{\text{in}}$ and $|\psi\rangle_{\text{out}}$ must be related via a unitary transformation. We can write this unitary in the form of a rotation matrix, e.g.

$$|\psi\rangle_{\text{out}} = R_\phi^\theta |\psi\rangle_{\text{in}}, \tag{4.6}$$

where

$$R_\phi^\theta = \begin{pmatrix} \cos\dfrac{\theta}{2} & -ie^{-i\phi}\sin\dfrac{\theta}{2} \\ -ie^{i\phi}\sin\dfrac{\theta}{2} & \cos\dfrac{\theta}{2} \end{pmatrix}. \tag{4.7}$$

The parameters θ and ϕ are related to the reflectivity at the interface and the phase shifts on reflection and transmission, and hence indirectly to the refractive indices of the optical media used to construct the interface. If the intensity reflection coefficient is $\mathcal{R}$ then for the state labelling used figure 4.9(a), where the transmitted modes retain the same label, we must have $\sin^2\frac{\theta}{2} = \mathcal{R}$ and $\cos^2\frac{\theta}{2} = 1 - \mathcal{R}$, i.e.

$$R_\phi^\theta = \begin{pmatrix} \sqrt{1-\mathcal{R}} & -ie^{-i\phi}\sqrt{\mathcal{R}} \\ -ie^{i\phi}\sqrt{\mathcal{R}} & \sqrt{1-\mathcal{R}} \end{pmatrix}.$$

For a 50:50 beam splitter with $\mathcal{R} = 0.5$ and with $\phi = \pi/2$ we obtain

$$R_y^{\pi/2} = \frac{1}{\sqrt{2}}\begin{pmatrix} 1 & -1 \\ 1 & 1 \end{pmatrix}.$$

Next, we want to show how this is equivalent to a rotation about the σ_y- axis in the Bloch sphere picture.

4.10 Rotations

In the previous section we saw an example where the change of state of a photon can be represented using a rotation matrix. In general for any two-state system, any change of state can be represented as a rotation of the Bloch (or Poincaré) vector. In quantum computing, these rotations are referred to as *single-qubit rotations* or *single-qubit gates*. For atoms (or other matter qubits) these rotations are implemented using external laser or microwave fields. For photons, rotations can be performed using polarizing media or beam splitters. For dual-rail encoded photons, e.g. using the scheme shown in figure 4.1, rotations can be implemented using *beam splitters*. In this section, first we shall consider the mathematical of rotation and then describe beam splitter examples.

In two-state quantum mechanics any operator, such as a rotation operator, can be written as a superposition of Pauli spin matrices. Using the Pauli spin basis it becomes easier to appreciate the equivalence of all two-state quantum systems from atoms to photons.

A general rotation matrix (see appendix A.3 for derivation),

$$R_{\hat{n}}^{\Theta} = \begin{bmatrix} \cos\dfrac{\Theta}{2} - in_z \sin\dfrac{\Theta}{2} & (-in_x - n_y)\sin\dfrac{\Theta}{2} \\ (-in_x + n_y)\sin\dfrac{\Theta}{2} & \cos\dfrac{\Theta}{2} + in_z \sin\dfrac{\Theta}{2} \end{bmatrix}, \tag{4.8}$$

rotates the state vector by an angle Θ about an axis given by the unit vector $\hat{n}$. For $\hat{n} = (0, 1, 0)$ and $\phi = \pi/2$ this gives the beam splitter matrix, equation (4.9). A diagramatic representation of the beam splitter rotation using both the Bloch sphere and the density matrix is show in figure 4.10.

4.11 Waveguide beam splitter

In this section, we consider the wave guide beam splitter shown in figure 4.9(b). We show that in this case the transformation is equivalent to a **Hadamard operator** which can also be represented using a rotation matrix. In chip-based photonics, a beam splitter is formed by bringing the two wave guide tracks closer together. This is illustrated in figure 4.11. When the two wave guides become close the fields leak across the gap, which allows the photon to tunnel back on forth. To model the hopping process we can write the photonic wavefunction as

$$|\psi\rangle = \frac{1}{\sqrt{2}}(e^{in_+kz}|+\rangle + e^{in_-kz}|-\rangle),$$

where $|\pm\rangle = \frac{1}{\sqrt{2}}(|0\rangle + |1\rangle)$ are the symmetric and anti-symmetric superpositions of the photon being in the left ('0') or right ('1') channels. The refractive indices,

$$n_{\pm} = \int P_{\pm}(x)n(x)dx,$$

where $P_{\pm}(x)$ is the probability distribution for the symmetric and anti-symmetric modes. A cross-section through the refractive index profile, $n(x)$, is shown in blue in figure 4.11. The symmetric superposition (red), the anti-symmetric superposition (green) and the total intensity are shown below. As the light propagates hopping between the two channels arises due to the phase difference between the two superpositions.

If the spacing is chosen correctly we obtain the equivalent of a 50:50 beam splitter. In figure 4.11, we show the wave guide tracks and the photon intensity as light propagates upwards. Each row shows an image at successive times for a

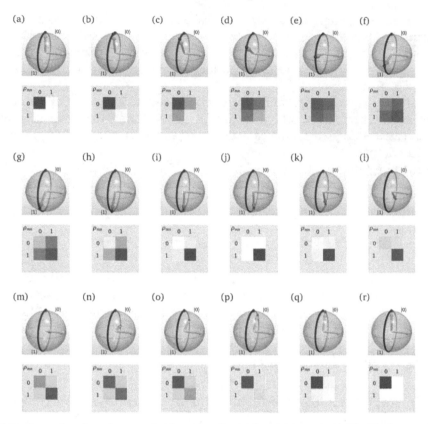

Figure 4.10. A rotation about the σ_y- axis (shown in yellow). The initial state is $|0\rangle$. The Bloch vector in blue. The final frame (r) corresponds to a rotation of $\Theta = 2\pi$. Note that the rotation direction follows a right-hand rule. The corresponding density matrix is shown below. In this case the rotation matrix is real, so if we begin with a state with real coefficient they remain real. Hence we only see red and green in the density matrix colormaps. An interactive figure is available at http://doi.org/10.1088/978-0-7503-2628-5.

particular minimum wave guide separation. As we move down the rows, the separation is progressively reduced. A cross-section of the refractive index profile and intensity at the peak intensity position (indicated by the yellow line) is shown below the main plot. The figure illustrates how it is possible to tune the reflectivity simply by varying the minimum distance between the two channels. In the second row, figures 4.11(g)–(l) (in the static figure version), we obtain a 50:50 beam splitter.

For the wave guide case, it makes more sense to use a labelling where effectively it is the reflected light that continues in the same mode. The unitary operator for a wave guide beam splitter, figure 4.12, with reflectivity $\mathcal{R}$ is

$$U_{\mathcal{R}} = \begin{pmatrix} \sqrt{\mathcal{R}} & i\sqrt{1-\mathcal{R}} \\ i\sqrt{1-\mathcal{R}} & \sqrt{\mathcal{R}} \end{pmatrix}.$$

Note that unlike the bulk optics beam splitter—where an asymmetry is introduce by which the path is reflected from a higher index step—in this case the two paths are

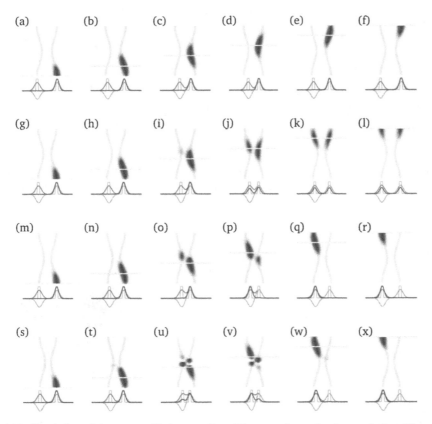

Figure 4.11. Simulation of the wave guide beam splitter. The rows shown the time evolution. The columns correspond to different wave guide spacings. For large spacing (a)–(f) there is no hopping. For (g)–(l) we see a 50:50 beam splitter (Hadamard). For (m)–(r) we hop once (equivalent to a NOT gate). For (s)–(x) we hop three times. An interactive figure is available at http://doi.org/10.1088/978-0-7503-2628-5.

identical. We can break this symmetry and remove the factors of i by considering that case where mode 1 is slightly shorter so picks up a phase $e^{-i\pi/2} = -i$ on both the input and output. This is illustrated schematically in figure 4.12(a). For a 50:50 beam splitter with $\mathcal{R} = 0.5$ including this path difference leads to the matrix,

$$H = \begin{pmatrix} 1 & 0 \\ 0 & -i \end{pmatrix} \frac{1}{\sqrt{2}} \begin{pmatrix} 1 & i \\ i & 1 \end{pmatrix} \begin{pmatrix} 1 & 0 \\ 0 & -i \end{pmatrix}, \tag{4.9}$$

$$= \frac{1}{\sqrt{2}} \begin{pmatrix} 1 & 1 \\ 1 & -1 \end{pmatrix}. \tag{4.10}$$

In computing, this matrix is known as the *Hadamard* operator. A Hadamard rotation in the Bloch sphere and density matrix representation is illustrated in figure 4.13. In the next section 4.14, we shall show how to obtain the same rotation for a two-level atom.

In all-optical quantum circuits, the wave guide beam splitter is often represented in a simpler form as in figure 4.12(b). We shall use this example as the key building

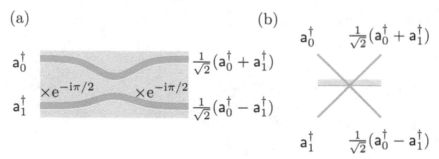

Figure 4.12. (a) Photonic wave guide beam splitter. The transfer matrix depends on the relative phase shifts in each channel and the phase shift induced by the wave guide coupling. For the case shown the matrix is given by equation (4.10). (b) Simplified schematic of a 50:50 wave guide beam splitter.

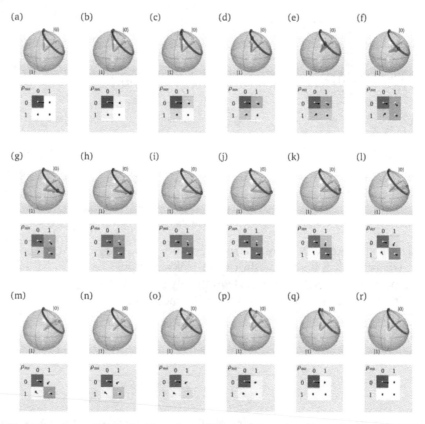

Figure 4.13. Trajectory on the Bloch sphere for a Hadamard operation. For a dual-rail encoded photon, section 4.9, a Hadamard is using a beam splitter like in figure 4.12. For a two-level matter qubit, section 4.14, the Hadamard is implemented using an external electromagnetic field from a laser or microwave source. An interactive figure is available at http://doi.org/10.1088/978-0-7503-2628-5.

block for all-optical quantum circuits such as single-photon interferometers and photonic two-qubit gates.

4.12 Photon creation: a single-photon emitter

So far, we have talked about single photons but not how they might be produced. In this section we shall discuss photon emission where an excitation of a matter qubit is converted into a photon or photonic qubit. We can re-use all the two-state quantum mechanics that we used to describe a beam splitter.

A source of individual photons is known as a *single-photon emitter*. Ideally, photon emitters produce single photons on demand, i.e. we would like to press a button and a photon comes out. This photon must have a well-defined mode such that successive photons are all identical. In addition, we would like to make copies of our single-photon emitter that each produce identical photons. Experimentally, this remains challenging. There are many types of single-photon emitters based on ions, single isolated atoms, defects in diamonds, quantum dots in semiconductors. For now, we only need to know that all photon emitters share the same basic ingredients. They have at least two energy levels. We can use an external field to excite the upper energy level. When the excited electron makes a transition from the upper energy level to the lower energy level, a photon is emitted. If we just take a single emitter in free space, the photons are emitted in many different directions, and only a small fraction can be collected by a lens. This directionality problem may be solved by putting the emitter inside a cavity (or wave guide). The cavity defines a spatial mode for the emitted photon. If the coupling between the cavity and the emitter is sufficiently strong then most photons are emitted into the cavity mode, and subsequently emerge in a particular direction. This is illustrated schematically in figure 4.14.

In the following sections, we shall focus on the mathematical description of photon creation. In particular, the excitation of a two-level system followed by decay. As we have emphasized, the mathematics of two-state quantum systems is equally applicable to any qubit system whether a two-level emitter or the two wave guides in a beam splitter.

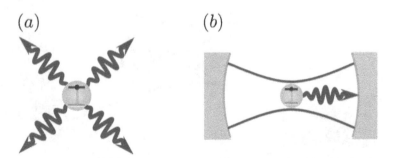

Figure 4.14. A two-level system. (a) Decay from the upper to the lower state leads to the emission of a photon in a random direction. (b) By placing the emitter into a cavity we can force the emission into the cavity mode.

4.13 Mathematical description of light–matter interactions

In this section, we shall develop a mathematical description of a two-state quantum emitter. As a starting point, we shall assume that we know everything about our two-state quantum system, and that it is completely decoupled from the environment. The interaction of a single two-level quantum system in a cavity with a single mode of the electromagnetic field is described by the **Jaynes–Cummings (JC) model** [14]. The model is applicable to the case of a single emitter in a cavity as in figure 4.15. The cavity ensures that the emitter predominantly couples to only one electromagnetic field mode.

The Jaynes–Cummings Hamiltonian can be written as

$$\mathcal{H} = \hbar\omega_0 \left(\hat{a}^\dagger \hat{a} + \frac{1}{2}\right) + \frac{1}{2}\hbar\omega_0 \hat{\sigma}_x + \frac{1}{2}\hbar g(\hat{a}^\dagger \hat{\sigma}^- + \hat{\sigma}^+ \hat{a}), \qquad (4.11)$$

where $\hat{a}^\dagger$ and $\hat{a}$ are the creation and annihilation operators for a photon in the mode of the cavity, $\hat{\sigma}^+$ and $\hat{\sigma}^-$ are the atomic raising and lower operators. The first and second terms are the energy of the field and the atom. The third term is light–atom interaction: $\hat{a}^\dagger \hat{\sigma}^-$ corresponds to emission, create a photon and transfer the atom from the upper to the lower level; $\hat{\sigma}^+ \hat{a}$ corresponds to the absorption of a photon, annihilation of a photon and transfer of the atom from the lower to the upper level. The parameter $g = (d\mathcal{E}/\hbar)\sqrt{\hbar\omega/2\epsilon_0 V}$, where V is the volume of the photon mode, is known as single-photon Rabi frequency, and measures the strength of the coupling between the atom and the cavity. There is a lot either hidden or missing in equation (4.11). Firstly, both the atomic and photonic operators are time dependent. Second, this Hamiltonian does not include any information about the loss of energy from the system, either in the form of a photon leaving the cavity or the atom decaying into a mode other than the cavity mode. Consequently, the JC model is only applicable when coherent dynamics dominates over loss. This is known as the *strong-coupling regime*. We shall return to this case when we discuss many photons in Part II. For

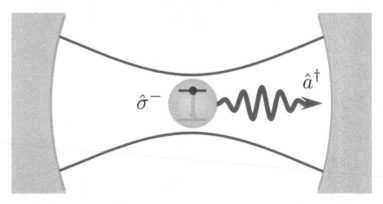

Figure 4.15. A two-level system inside in a cavity. The coupling constant between the atom and the spatial mode of the cavity is known as the single-photon Rabi frequency, $g = (d\mathcal{E}/\hbar)\sqrt{\hbar\omega/2\epsilon_0 V}$, where V is the volume of the cavity mode. The transfer of excitation from the atom to cavity is described by the operator product of the atomic lowering operator $\hat{\sigma}^-$ and the photon creation operator $\hat{a}^\dagger$.

now we shall restrict ourselves to the semi-classical limit where we can replace the photon operators with a constant.

The Rabi model. In the limit of many photons, or when the atom is unable to resolve the difference between n and $n + 1$ photons, the JC model becomes equivalent to the **Rabi model**, where the electromagnetic field can be treated classically. In this **semi-classical limit** (matter quantized, field not), the interaction part of the Hamiltonian becomes,

$$\mathcal{H}_{int} = \frac{1}{2}\hbar\Omega(\hat{\sigma}^- + \hat{\sigma}^+) = \frac{1}{2}\hbar\Omega\hat{\sigma}_x, \tag{4.12}$$

where $\Omega = g\bar{n}$ and $\bar{n}$ is the number of photons in the mode, and $\hat{\sigma}_x$ is the x component of the Pauli matrices. This is known as the **Rabi model**. We shall use this model to describe the excitation of our two-level single-photon emitter using external fields. This excitation step is a precursor to single-photon emission. Over the next few sections, we shall look at the time dependence in this model, and how the model can be extended to include decoherence arising from the interaction between our quantum emitter and the environement.

4.14 Two-level system driven by a near-resonant field

In this section, we consider how to excite (or de-excite) any two-level system in an atom, ion, molecule, artificial atom, etc. In the context of quantum optics this section is particularly useful in developing our understanding of atom–photon interactions, for example, in the case of single-photon emitters. We assume that we have two energy levels $|g\rangle$ and $|e\rangle$ with energies E_0 and E_1. As we are dealing with a two-state system, we can use the same mathematical toolkit as before, only replacing the labels $|0\rangle$ and $|1\rangle$ with $|g\rangle$ and $|e\rangle$.

We assume that in order to induce a rotation of the state vector, the atom is subject to an EM field with electric field component, $\mathcal{E} = \mathcal{E}_0 \cos(\phi_L - \omega t)$, where ϕ_L is the phase and ω is angular frequency of the light. The difference between the angular frequency of the light, ω, and the atomic resonance, $\omega_0 = (E_1 - E_0)/\hbar$, is known as the *detuning*, $\Delta = \omega - \omega_0$. The energy levels and incident field are shown schematically in figure 4.16.

To make a single-photon emitter, we first need to drive the emitter into the excited state. Typically, we would do this using a laser or microwave. We assume that this external field has many photons per mode and we can treat it classically. The electric field of the light can be written as $\mathcal{E} = \mathcal{E}_0 \cos(\phi_L - \omega t)$, where ϕ_L is the phase and ω is angular frequency of the light.

The external field induces a superposition of the two states, $|g\rangle$ and $|e\rangle$. As the two states has different charge distributions, the superposition corresponds to oscillation charge and hence an induced electric dipole. The coupling between the light and the atom is given by a Hamiltonian $\mathcal{H}' = -\mathbf{d} \cdot \mathcal{E}$, where $\mathbf{d} = -e\mathbf{r}$ is the electric dipole operator. To understand the dynamics of the atom in the light field we need to solve the time-dependent Schrödinger equation,

$$i\hbar\partial_t|\psi\rangle = (\mathcal{H}_0 + \mathcal{H}')|\psi\rangle,$$

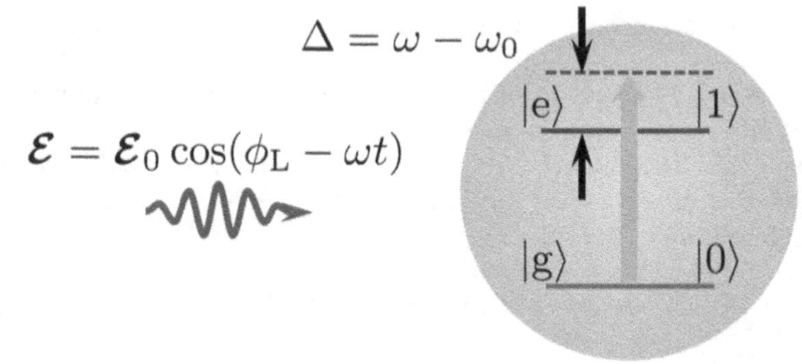

$$\Delta = \omega - \omega_0$$

$$\mathcal{E} = \mathcal{E}_0 \cos(\phi_{\mathrm{L}} - \omega t)$$

Figure 4.16. Schematic of a two-state quantum system. A change of state is induced by an external field, for example a classical laser field $\mathcal{E} = \mathcal{E}_0 \cos(\phi_{\mathrm{L}} - \omega t)$, with phase ϕ_{L} and angular frequency ω. For a single-photon emitter we assume that the excited state $|e\rangle$ decays emitting a photon.

where $\mathcal{H}_0$ is the Hamiltonian for the quantum system and $\mathcal{H}'$ is the perturbation due to the EM field. Substituting a state vector of the form

$$|\psi(t)\rangle = a(t)|g\rangle e^{-iE_0 t/\hbar} + b(t)|e\rangle e^{-iE_1 t/\hbar},$$

we find (see appendix A.4)

$$|\psi(t)\rangle = \mathsf{R}_{\hat{n}}(\Theta)|\psi(0)\rangle,$$

where

$$\mathsf{R}_{\hat{n}}(\Theta) = \begin{bmatrix} \cos\dfrac{\Theta}{2} - i\dfrac{\Delta}{\Omega_{\mathrm{eff}}}\sin\dfrac{\Theta}{2} & -i\dfrac{\Omega}{\Omega_{\mathrm{eff}}}e^{-i\phi_{\mathrm{L}}}\sin\dfrac{\Theta}{2} \\[2ex] -i\dfrac{\Omega}{\Omega_{\mathrm{eff}}}e^{i\phi_{\mathrm{L}}}\sin\dfrac{\Theta}{2} & \cos\dfrac{\Theta}{2} + i\dfrac{\Delta}{\Omega_{\mathrm{eff}}}\sin\dfrac{\Theta}{2} \end{bmatrix},$$

This is known as the *Rabi solution* and is analogous to the rotation matrices such as equation (4.8). By choosing the parameters, Ω, Δ, ϕ_{L} and t we can perform any desired rotation. For example, using $\Omega = \Delta$, $\phi_{\mathrm{L}} = \pi/2$ and $\Omega_{\mathrm{eff}} t = \pi$ we can implement a Hadamard rotation identical to the beam splitter case characterized by equation (4.10), see figure 4.13. In figure 4.17 we show another example, which implements a cyclic rotation from $|0\rangle$ to $|+\rangle_x$ to $|+\rangle_y$ and back to $|0\rangle \equiv |+\rangle_z$. The Rabi solution provides a useful description of how to excite our single-photon emitter and hence implement photon creation.

4.15 Decay and decoherence

A two-level emitter is unlikely to be perfect because in practice excited states decay spontaneously in a random direction and the emitter is constantly being perturbed by its environment. To model these effects it is necessary to extend the basic two-state model to include decay or environmentaly induced decoherence. The concept of decay is quite wide ranging and could arise due to many physical processes. For example, if our state corresponds to a photon with a particular spatial mode then if some process distorts the mode then amplitude of the state is decreased.

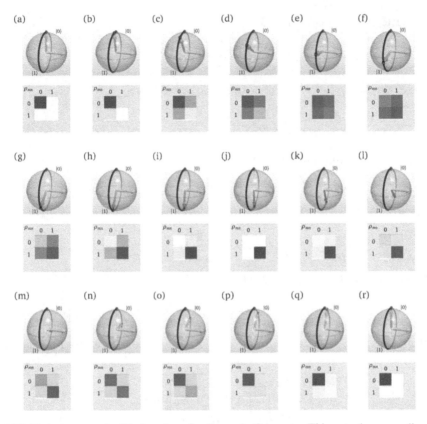

Figure 4.17. Trajectory on the Bloch sphere for the cycle C_{xyz} gate. This gate does a cyclic rotation $|0\rangle \rightarrow |+\rangle_x \rightarrow |+\rangle_y$. An interactive figure is available at http://doi.org/10.1088/978-0-7503-2628-5.

Another type of decay arises when our quantum two-state system couples to the environment. This type of decay is referred to as *decoherence*. In this section, we introduce decay into the two-state model in the form of an ensemble average. This is in contrast to chapter 5, where we considered individual trajectories.

Derivation of the optical Bloch equations. First, we derive the equations of motion for the components of the Bloch vector for a pure state (no decoherence). The equation of motion of an expectation value of any operator can be derived with the Schrödinger equation. For example, starting from

$$\frac{\partial}{\partial t}\langle\psi|\sigma|\psi\rangle = \frac{\partial\langle\psi|}{\partial t}\sigma|\psi\rangle + \langle\psi|\sigma\frac{\partial|\psi\rangle}{\partial t},$$

and then using the Schrödinger equation and its complex conjugate,

$$\frac{\partial}{\partial t}|\psi\rangle = -\frac{i}{\hbar}\mathcal{H}_{int}|\psi\rangle \quad \text{and} \quad \frac{\partial}{\partial t}\langle\psi| = \frac{i}{\hbar}\langle\psi|\mathcal{H}_{int},$$

where we have used the Hermitian property of the Hamiltonian, we obtain

$$\frac{\partial}{\partial t}\langle\psi|\sigma|\psi\rangle = \frac{i}{\hbar}(\langle\psi|\mathcal{H}_{int}\sigma|\psi\rangle - \langle\psi|\sigma\mathcal{H}_{int}|\psi\rangle).$$

We can write this as a commutator

$$\frac{\partial}{\partial t}\langle \sigma \rangle = \frac{i}{\hbar}\langle [\mathcal{H}_{\text{int}}, \sigma] \rangle.$$

If we take $\mathcal{H}_{\text{int}} = (\hbar/2)(\Delta \sigma_z + \Omega \sigma_x)$, then for σ_x we have

$$\frac{\partial}{\partial t}\langle \sigma_x \rangle = \frac{i}{2}\langle [\Delta \sigma_z + \Omega \sigma_x, \sigma_x] \rangle.$$

Next, we use the commutation relations

$$[\sigma_x, \sigma_y] = 2i\sigma_z, \quad [\sigma_z, \sigma_x] = 2i\sigma_y, \quad [\sigma_y, \sigma_z] = 2i\sigma_x,$$

and $[\sigma_j, \sigma_j] = 0$, which gives

$$\frac{\partial}{\partial t}\langle \sigma_x \rangle = \frac{i}{2}\Delta \langle [\sigma_z, \sigma_x] \rangle = \frac{i}{2}\Delta \langle 2i\sigma_y \rangle = -\Delta \langle \sigma_y \rangle.$$

Using $u = \langle \sigma_x \rangle$ and $v = \langle \sigma_y \rangle$ this becomes

$$\frac{\partial}{\partial t}u = -\Delta v.$$

Repeating this derivation for the y and z components we obtain the coupled equations

$$\begin{pmatrix} \dot{u} \\ \dot{v} \\ \dot{w} \end{pmatrix} = \begin{pmatrix} 0 & -\Delta & 0 \\ \Delta & 0 & -\Omega \\ 0 & \Omega & 0 \end{pmatrix} \begin{pmatrix} u \\ v \\ w \end{pmatrix},$$

where $u = \langle \sigma_x \rangle$, $v = \langle \sigma_y \rangle$, and $w = \langle \sigma_z \rangle$. Classically, we can think of this as a torque $(\Omega, 0, \Delta)$ acting on the Bloch vector (u, v, w)[6].

Decoherence may disrupt both the relative phase of the $|0\rangle$ and $|1\rangle$ components (encoded in u and v) and their amplitude (encoded in w). If we assume that the 'excited' state $|1\rangle$ decays at a rate $\Gamma = 1/T_1$ ($\dot{\rho}_{bb} = -(1/T_1)\rho_{bb}$, $\dot{\rho}_{aa} = (1/T_1)\rho_{bb}$, therefore $\dot{w} = \dot{\rho}_{aa} - \dot{\rho}_{bb} = 2(1/T_1)\rho_{bb} = -(1/T_1)(w - 1)$). If this decay plus any other interactions that perturb the relative phase cause u and v to decay at a rate $1/T_2$ then the equations of motion become

$$\dot{u} = -\frac{u}{T_2} - \Delta v, \tag{4.13}$$

$$\dot{v} = \Delta u - \frac{v}{T_2} - \Omega w, \tag{4.14}$$

$$\dot{w} = \Omega v - \frac{1}{T_1}(w - 1). \tag{4.15}$$

---

[6] As noted earlier, historically atomic physics texts used a difference definition of u, v and w. w is sometimes defined with the opposite sign which flips the Bloch sphere such that $|1\rangle$ is at the North pole. This also changes the sign of v such that Δ replaced by $-\Delta$, and changes the form of any damping terms.

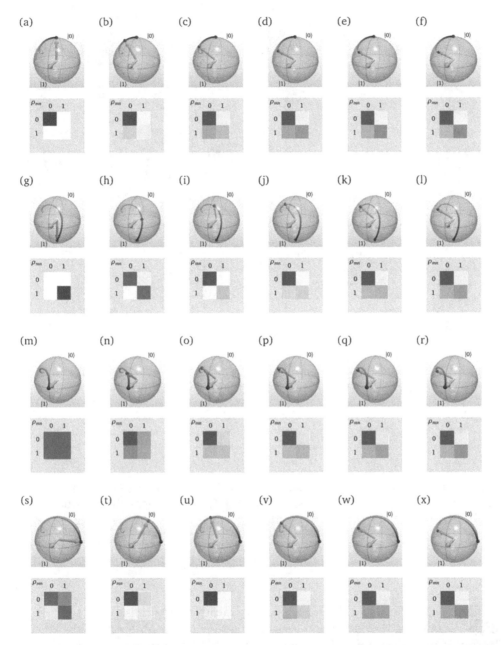

Figure 4.18. Solution of the optical Bloch equation as a function of time t and Rabi frequency, Ω. An interactive figure is available at http://doi.org/10.1088/978-0-7503-2628-5.

These equations are known as the **optical Bloch equations**.

Next, we show two example solutions of the optical Bloch equations for different parameters, figures 4.18 and 4.19. First, in figure 4.18, we consider the case of isolated qubits, where the only decay mechanism is radiative (i.e. spontaneous

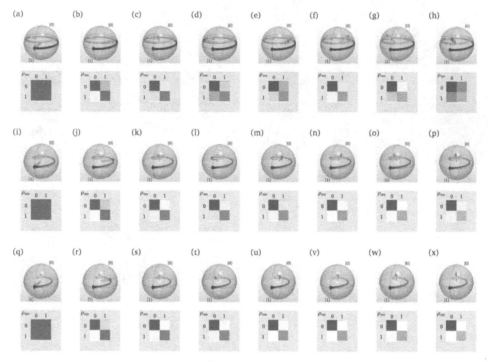

Figure 4.19. Evolution of the Bloch vector and density matrix for a two-level system initial prepared in the superposition state $|\psi\rangle = \frac{1}{\sqrt{2}}(|0\rangle + |1\rangle)$. The Rabi frequency and detuning are $\Omega = 0$ (no driving) and $\Delta = 20\Gamma$. The columns show the time evolution. The different rows correspond to increasing decoherence, (a)–(h) $T_2 = 2T_1$, (i)–(p) $T_2 = T_1/3$ and (q)–(x) $T_2 = 2T_1/11$, respectively. Note how in (x) the coherences have decayed almost completely. The Bloch vector sits on the vertical axis and the off-diagonal elements in the density matrix are white (zero). An interactive figure is available at http://doi.org/10.1088/978-0-7503-2628-5.

emission) In this case, we can write $1/T_1 = \Gamma$ and $1/T_2 = \Gamma/2$, where Γ is the spontaneous decay rate of the excited state ($|1\rangle$). The optical Bloch equations for spontaneous decay are

$$\dot{u} = -(\Gamma/2)u - \Delta v,$$
$$\dot{v} = \Delta u - (\Gamma/2)v - \Omega w,$$
$$\dot{w} = \Omega v - \Gamma(w - 1).$$

Figure 4.18 illustrates the trajectories for steady-state driving. In each case, the drive parameters, Ω and Δ are the same. Only the initial condition is changed. We learn two important points from this figure. First, that it does not matter where we start (what initial condition), we always end up with the same steady-state. This is a property of the *non-Markovian* nature of decay—all memory of the initial state is erased. The second important point hidden in figure 4.18, is that the steady-state values of the coherences—the off-diagonal elements in the density matrix—are non-zero. The balance between a coherent drive and incoherent drive means that coherence survives for continuous driving.

The steady-state solutions of the optical Bloch equations are particularly useful in the description of the initialization of atomic qubits, e.g. the mathematical description of laser cooling of atoms and ions, and laser trapping of atoms. The steady-state solution is found by setting the time variation to zero.

$$\dot{u} = -(\Gamma/2)u - \Delta v = 0,$$
$$\dot{v} = \Delta u - (\Gamma/2)v - \Omega w = 0,$$
$$\dot{w} = \Omega v - \Gamma(w - 1) = 0.$$

From the first and third equation, we have

$$u = -\frac{2\Delta}{\Gamma}v \quad \text{and} \quad v = \frac{\Gamma}{\Omega}(w - 1).$$

Substituting into the second equation gives

$$-\frac{2\Delta^2}{\Gamma}\frac{\Gamma}{\Omega}(w - 1) - \frac{\Gamma^2}{2\Omega}(w - 1) - \Omega w = 0.$$

Rearranging we find the population inversion is

$$w = \frac{4\Delta^2 + \Gamma^2}{2\Omega^2 + 4\Delta^2 + \Gamma^2} = \frac{1}{1 + s},$$

where

$$s = \frac{\Omega^2/2}{\Delta^2 + \Gamma^2/4},$$

is known as the saturation parameter. The steady-state probability to be in state $|1\rangle$ is given by

$$P_{|1\rangle} = \frac{1}{2}(1 - w) = \frac{\Omega^2/4}{\Delta^2 + \Gamma^2/4 + \Omega^2/2}.$$

The transition to this steady-state (regardless of input state) is illustrated in figure 4.18.

In the second example, figure 4.19, we show what happens when the coherences decay faster than the population inversion. Consider the case, $T_1 \to \infty$, i.e. the excited state does not decay over the time scale of interest, and there is no driving, $\Omega = 0$. In this case, if we begin with the Bloch vector along the x axis, $(u, v, w) = (1, 0, 0)$, we obtain[7] $u(t) = e^{-t/T_2}\cos\Delta t$ and $v(t) = e^{-t/T_2}\sin\Delta t$, i.e. the

---

[7] If $\Omega = 0$ then adding $\dot{u}$ and $i\dot{v}$ gives,

$$\dot{u} + i\dot{v} = -\left(\frac{1}{T_2} - i\Delta\right)(u + iv).$$

We can integrate this to find that

$$u(t) + iv(t) = e^{-(1/T_2 - i\Delta)t}[u(0) + iv(0)].$$

Inserting the initial condition and equating real and imaging parts, we obtain $u(t)$ and $v(t)$.

Bloch vector spirals in towards the centre of the Bloch sphere, see figure 4.19(left). The density matrix is given by,

$$\rho = \frac{1}{2}\begin{pmatrix} 1 & e^{-t/T_2}e^{-i\Delta t} \\ e^{-t/T_2}e^{i\Delta t} & 1 \end{pmatrix}.$$

The off-diagonal terms—the coherences—decay with a time constant T_2. The decay of the coherences is known as *decoherence*. Decoherence may arise due to the interaction of the two-level system with the environment. In figure 4.19 we show the trajectory on the Bloch shpere and the corresponding density matrix for a case where $T_2 \ll T_1$. We see that the effect of decoherence is for the Bloch vector to collapse towards the vertical axis. Or in the density matrix visualization, the off-diagonal squares turn white, corresponding to zeros.

The optical Bloch equations model the effect of decay and decoherence on an ensemble of identical two-state quantum systems. But they do not tell us anything about the individual trajectories for which we have to use a useful mathematical description of individual trajectories known as the *quantum jump method* (section 5.1).

4.16 Detecting single photons: homodyne detection

How do we know we have a single photon? As Feynman said, now that we have sensitive detectors like photomulitpliers and avalanche photodiodes we know we have a photon because our detector makes a click. But a detector click is not a sufficient proof of the existence of photons. A much better way to see photons is **homodyne detection**. In a homodyne detector the quantum field is interfered with a classical field. A basic set up for homodyne detection is shown in figure 4.20. The quantum field to be measured—represented by the state vector $|\psi\rangle$—is interfered with a classical field—called the **local oscillator**—that is represented by $|\alpha\rangle$. The

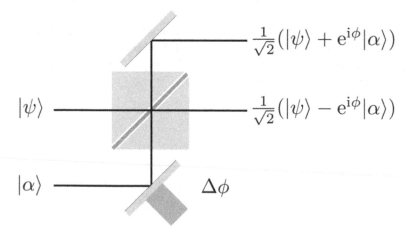

Figure 4.20. Homodyne detection: the quantum light field, labelled by the state vector $|\psi\rangle$ is interfered with a classical local oscillator labelled $|\alpha\rangle$ where α is proportional to the amplitude of the field. The difference between the two output intensities is proportional $|\langle\alpha|\cos\phi|\psi\rangle|$.

difference between the two output ports is proportional to $|\langle\alpha|\cos\phi|\psi\rangle|$. Consequently, we can use the local oscillator field, with amplitude proportional to $|\alpha|$ to amplify the quantum field signal. By varying the phase of a local oscillator, the electric field of the quantum field at different relative phase is measured. This is equivalent to measuring the electric field in space for fixed time, or as a function of time at a particular position.

Simulated electric field values for different types of photonic states are plotted in figure 4.21. The columns show different values of the relative phase. The rows show different photon numbers distributions.

Wigner representation. Although the natural basis to talk about photon number is the **Fock state basis**, frequently we might have a state of light that consists of a superposition of Fock states. We could represent this superposition using a density matrix in the photon number (Fock) state basis. Alternatively, we can use a phase space representation such as the **Wigner function**. Next, we aim to make a connection between these different representations of the electromagnetic field. In figure 4.21 the Wigner function is shown inset.

The Wigner function, W, is a phase space representation. For a one-dimensional mechanical harmonic oscillator, phase space is a function of momentum, p, and position, q, respectively. The Wigner function is typically plotted in the complex

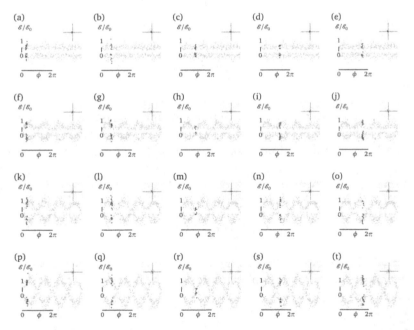

Figure 4.21. Homodyne detection of the electric field $\mathcal{E}$ as a function of the phase of the local oscillator, ϕ. Columns show a selected range of ϕ indicated in red. The rows show different states of light. Panels (a)–(e) are the vacuum state. Panels (f)–(j) are a single photon. Panels (k)–(y) show superpositions of photon numbers, n, with intensity P_n indicated by the histograms inset lower right. The corresponding Wigner function is shown as colour map, inset upper right. Green is positive and purple is negative. Selected a particular value of ϕ can be used to reconstruct the time evolution of the state. An interactive figure is available at http://doi.org/10.1088/978-0-7503-2628-5.

plane, as a function of the complex parameter, $z = p + iq$. For light, the axes of phase space correspond to the quadratures of the electric field, $\mathcal{E}$. In complex representation, the two quadratures are the real and imaginary parts of $\mathcal{E}$. We shall use the p and q labels for brevity.

The Wigner function is often referred to as a **quasi-probability distribution**. It provides information about the probability of measuring particular values of the field. However, as for a quantum state, the Wigner function can be negative it is not a real probability distribution. Wigner function negativity is sometimes used as a measure of quantumness or non-classicality. However, there are classes of non-classical states that do not display any negativity, such as squeezed states, see Part II. Experimentally, the Wigner function cannot be measured directly, but can be reconstructed from data obtained using homodyne detection. The field and corresponding Wigner functions are shown in figure 4.21.

Fock states: the vacuum and single-photon state. The top two rows in figure 4.21(inset) show the Wigner function for the $|n\rangle = |0\rangle$ (the vaccum state) and $|1\rangle$ (single photon) Fock states. The vacuum state and single-photon Wigner functions are identical to the ground state and first excited state of 1D harmonic oscillator. Adding more quanta is equivalent to moving up the ladder of harmomic oscillator states. For all Fock states, the Wigner function has circular symmetry. There is no phase or time dependence. This bizarre property of photon number states is sometimes discussed as related to number-phase indeterminacy[8]. The concept of the phase of a photon is tricky and we shall need to revisit it. As electric field oscillates so fast, field amplitudes are not directly observable, so often we cannot know or say anything useful about them. However, the concept of relative phase which is defined by the geometry of an interference experiment is useful. We could say that a Fock state has the property of being a superposition of all possible phases at once. However, when we split a Fock state on a 50/50 beam splitter, the two output modes are phase-related and we will see interferences in a homodyne measurement. Similarly, if we allow the photon to follow different paths as in Young's double-slit experiment, the two paths have a well-defined relative phase, and we can see interference.

The Wigner function of the vacuum state—top row in figure 4.21 is a circular Gaussian distribution,

$$\mathcal{W}_{|0\rangle}(z) = \frac{2}{\pi}e^{-|z|^2},$$

where $z = p + iq$. This is plotted as a colormap in the inset of figures 4.21(a)–(e). The Wigner function of a single photon is

$$\mathcal{W}_{|1\rangle}(z) = \frac{1}{\pi}(2q^2 + 2p^2 - 1)e^{-|z|^2}.$$

This is negative at the origin $\mathcal{W}_{|1\rangle}(0) = -1/\pi$. The single-photon Wigner function is shown inset in figures 4.21(f)–(j). As a function of phase (or time) the Wigner function rotates in the complex plane,

---

[8] For a theoretical derivation see, e.g. [15].

$$\mathcal{W}(p, q, t) = \mathcal{W}(q \cos \omega t - p \sin \omega t, \ q \sin \omega t + P \cos \omega t, \ 0).$$

The electric field is given by the projection of the Wigner function onto the real axis,

$$\frac{\mathcal{E}}{\mathcal{E}_0} = \int_{-\infty}^{\infty} \mathcal{W}(z) \mathrm{d}p.$$

Hence the amplitude of the electric field noise—or quantum noise—is related to the width of the Wigner function along the real axis.

If we take a distribution of photons numbers then their overall relative phases become important. In figures 4.21(k)–(o) we see that even a superposition of a vacuum state and only one photon, i.e. $|\psi\rangle = a|0\rangle + b|1\rangle$ where $|0\rangle$ and $|1\rangle$ are Fock states, has a time evolution that begins to look oscillatory. As we add more photons, the bottom two rows in figure 4.21, the Wigner function becomes more localized. The electric fields illustrated in figure 4.21 match what is observed experimentally, see e.g. [16].

Coherent states. A state with a Poissonian distribution of photon numbers is known as a **coherent state**. Coherent states were first introduced by Roy Glauber in 1963 to help explain the photon counting results of Hanbury Brown and Twiss in [17]. The coherent state concept proved very useful as it provides a convenient representation of laser fields. The coherent state, labelled $|\alpha\rangle$ has an amplitude $|\alpha|$. We can write

$$|\alpha\rangle = \sum_{n=0}^{\infty} c_n |n\rangle,$$

where the coefficients

$$c_n = \mathrm{e}^{-\alpha^2/2} \frac{\alpha^n}{\sqrt{n!}}.$$

Why is coherent state often thought of as being classical or semi-classical? A clue is given by looking at the Wigner function. The Wigner function of a coherent state is

$$\mathcal{W}_{|\alpha\rangle}(z) = \mathrm{e}^{-|z-\alpha|^2},$$

which is a Gaussian displaced by $|\alpha|$ from the origin, see figures 4.21(u)–(y). In time this looks like the sine wave we expect for a classical field. Hence coherent states are often used to model the classical state of light such as laser fields. It is worth noting that although in the standard definition of the coherent state, the relative phases between Fock states are all zero, this is not a requirement, as for 'classical' light observables are not sensitive to this property, see e.g. [18].

Equivalence with a charge moving in a harmonic potential. The coherent state vector is identical to a Poissonian superposition of harmonic oscillator states. In fact we can write any light state as a superposition of harmonic oscillator states. Consequently, there is a strong analogy between the motion of a particle moving in a harmonic potential and the states of light. A good example that has been exploited experimentally is the motion of an ion in an ion trap, see e.g. [19] and references therein. If we excite the ion into a Poissonian superposition of vibrational levels it becomes localized within a Gaussian wavepacket that oscillates back and forth. The motion of the ion in space mimics the electric field values observed for coherent states of light.

We can generate a state coherent by acting on the vacuum state using the displacement operator,

$$|\alpha\rangle = \hat{D}(\alpha)|0\rangle = e^{-|\alpha|^2/2} \sum_{n=0}^{\infty} \frac{\alpha^n}{\sqrt{n!}} |n\rangle.$$

where

$$\hat{D}(\alpha) = e^{-(\alpha \hat{a}^\dagger + \alpha * \hat{a})}.$$

To convert from the photon number distribution to an electric field we use the electric field operator. For a single-mode polarized along x, the electric field operator is

$$\hat{\mathcal{E}} = \mathcal{E}_1[\hat{a}(t) + \hat{a}^\dagger(t)]\cos kz,$$

where $\mathcal{E}_1 = \sqrt{\hbar \omega / \epsilon_0 V}$ is the electric field of one photon in a volume V[9], and the expectation value of the electric field is

$$\langle \alpha | \mathcal{E} | \alpha \rangle = \mathcal{E}_1[\alpha e^{-i\omega t} + \alpha * e^{i\omega t}] \cos kz,$$
$$= 2\mathcal{E}_1|\alpha|\cos kz.$$

The rms fluctuations in the electric field are,

$$\langle \Delta \hat{\mathcal{E}}^2(t) \rangle^{1/2} = \mathcal{E}_1| \cos kz|.$$

Note that the fluctuations are independent of the amplitude $|\alpha|$. So even, for zero field $\alpha = 0$ we observe fluctuations. These fluctuations are often referred to either as zero-point fluctuations or **quantum noise**.

References

[1] Einstein A 1931 Maxwell's influence on the development of the conception of physical reality *James Clerk Maxwell—a commemorative volume 1831–1931* (Cambridge: Cambridge University Press) (written for the centenary of Maxwell's birth)
[2] Roychoudhuri C, Kracklauer A F and Creath K 2008 *The Nature of Light: What is a Photon?* (Boca Raton: CRC Press)
[3] Feynman R P 1985 *QED: The Strange Theory of Light and Matter* (Princeton, NJ: Princeton University Press)

---

[9] This follows from integrating the energy density,

$$\int \frac{1}{2}\left(\epsilon_0 \mathcal{E}_1^2 + \mu_0 \mathcal{B}_1^2\right) = \frac{\hbar \omega}{V},$$

with $\mathcal{B}_1 = \mathcal{E}_1/c$ in free space.

[4] Breitenbach G, Schiller S and Mlynek J 1997 Measurement of the quantum states of squeezed light *Nature* **387** 471–5

[5] Lvovsky A I, Hansen H, Aichele T, Benson O, Mlynek J and Schiller S 2001 Quantum state reconstruction of the single-photon fock state *Phys. Rev. Lett.* **87** 050402

[6] Barnett S M, Babiker M and Padgett M J 2017 Optical orbital angular momentum *Phil. Trans. R. Soc.* A **375** 20150444

[7] von Neumann J 1927 Wahrscheinlichkeitstheoretischer Aufbau der Quantenmechanik *Nachrichten von der Gesellschaft der Wissenschaften zu Göttingen, Mathematisch-Physikalische Klasse* **1927** 245–72

[8] Adams C S and Hughes I G 2019 *Optics f2f: From Fourier to Fresnel* 1st edn (oxford University Press) https://doi.org/10.1093/oso/9780198786788.001.0001

[9] Poincaré H 1881 Mémoire sur les courbes définiés par une équation differentielle *J. Math.* **7** 375–422

[10] Poincaré H 1892 Sur l'Analysis situs *C. R. Acad. Sci.* **115** 633–6

[11] O'Brien J L 2007 Optical quantum computing *Science* **318** 1567–70

[12] Politi A, Cryan M J, Rarity J G, Yu S and O'Brien J L 2008 Silica-on-silicon waveguide quantum circuits *Science* **320** 646–9

[13] Hénault F 2015 Quantum physics and the beam splitter mystery *Proc. SPIE 9570, The Nature of Light: What are Photons? SPIE Optical Engineering + Applications,* VI *(San Diego, CA)* 95700Q

[14] Jaynes E T and Cummings F W 1963 Comparison of quantum and semiclassical radiation theories with application to the beam maser *Proc. IEEE* **51** 89–109

[15] Carruthers P and Nieto M M 1965 Coherent states and the number-phase uncertainty relation *Phys. Rev. Lett.* **14** 387

[16] Bellini M, Zavatta A and Viciani S 2005 From quantum to classical: watching a single photon become a wave *The Nature of Light: What is a Photon?* ed C Roychoudhuri, A F Kracklauer and K Creath (Boca Raton, FL: CRC Press)

[17] Hanbury Brown R and Twiss R Q 1956 Correlation between photons in two coherent beams of light *Nature* **177** 27–9

[18] Mølmer K 1997 Optical coherence: A convenient fiction *Physical Review A* **55** 3195–203

[19] Flühmann C and Home J P 2020 Direct characteristic-function tomography of quantum states of the trapped-ion motional oscillator *Phys. Rev. Lett.* **125** 043602

IOP Publishing

An Interactive Guide to Quantum Optics

Nikola Šibalić and C Stuart Adams

Chapter 5

Measurements: projective and non-destructive

Albert Einstein: *What is?*

Niels Bohr: *What can be said?*

The natural sciences were built on a tacit assumption: information about the Universe can be acquired without changing its state. The ideal of 'hard science' was to be objective and provide a description of reality. Information was regarded as unphysical, ethereal, a mere record of the tangible, material Universe, an inconsequential reflection, existing beyond and essentially decoupled from the domain governed by the laws of physics. This view is no longer tenable [1]

 LabBricks literature graph for this chapter can be found at https://labbricks.com/#8/qrA8Affq8Z

doi:10.1088/978-0-7503-2628-5ch5

A friend was visiting Niels Bohr in his home in Tisvilde when he noticed a horseshoe above Niels's doorway, 'Niels, why is that hanging there?'. 'Oh, it's for luck' replied Bohr. 'But surely you don't believe such superstitious nonsense', contested the friend. 'Oh no...', replied Bohr, '... I'm told that it works even if you don't believe it'. A horseshoe in this story[1] is a perfect allegory for a set of unquestioned beliefs, what in social sciences would be called ideology. One such unquestioned belief runs so deeply in the science of the enlightenment that it was not even thought of as an assumption, let alone written down and discussed. It is the idea that things happen in the world regardless of whether we look at them or not. This unquestioned property of the world we would name *reality*[2].

Trying to reconcile experimental observations and theoretical models, we could say that it seems that actually 'things happen in the world consistently with available information about them'. This sounds like a minor rephrasing, but it was standing in contrast to prevailing ideology of a time, and caused a great concern for physicists of the quantum age, succinctly expressed in Einstein's quip to Bohr 'Do you really believe that the Moon isn't there when nobody looks?'

This rephrasing allows space for two important things:

(i) a system does not have to behave the same way when we don't look at it and when we do, allowing, e.g. photons, to travel as waves but be detected always as particles; and

(ii) information available to us[3] about the system becomes important for dynamics of the system.

Suddenly measurement becomes much more than a snapshot of a system in a given time, and the process of retrieval of information from the system, and how that influences the system, requires a whole (important) chapter in our quantum story.

In the following, we will see how we can formulate a quantitative way to consistently track this new 'information view'. Then we will be ready to explore a gallery of examples observable in experiments demonstrating the two important aspects of quantum dynamics: (i) abrupt 'jumpiness' brought by strong measurement; and (ii) smooth coherent evolution modified by 'information leakage and overlap' through observation.

5.1 Bang, and what happens next: quantum jumps and quantum regression theorem

The aim of science is not to open the door to infinite wisdom, but to set a limit to infinite error.

Bertolt Brecht in 'Life of Galileo' (1943)

---

[1] The origin of the story is disputed, but that doesn't matter for the point.

[2] Interestingly, if we go further back in history, one could argue that the ancient Greek philosopher Plato did question this point with his Allegory of the Cave. In that story a group of people sit with their backs turned to the Sun that comes through the cave's entrance, watching projected shadows of the world on the cave wall facing them and believing that's what the world is. Until one brave soul turns around and sees the actual world.

[3] 'Information available to us' does not mean that we have necessarily measured with our instruments, but maybe something else recorded information and we could in principle retrieve information from that.

Zooming into a probability subspace. Let's start with classical information: imagine there is large number of balls in a box. They can be labeled with stickers A or B or both with stickers A and B. We know the probability that a ball is labeled with sticker A is P_A, probability that it has label B is P_B, and probability that a single ball has both labels is P_{AB}. We took one ball from the box, and for now all information about that ball can be summarized as these probabilities. But then we uncover under one of each fingers the label B on the ball. Now the information about the ball available to us is different: we now know that our ball lives in the subspace of balls that have the label B. Probability that now the ball also has the label A is also changed now, and is given by conditional probability $P_{A|B}$ (probability of A if B is present) as

$$P_{A|B} = \frac{P_{AB}}{P_B}.$$

We are essentially measuring the extent of space where both A and B are happening P_{AB} relative to the total subspace where B already happened P_B. If the situation is like in figure 5.1, detecting label B makes it very likely that our ball will also have label A. In the other case, if we first found label A on one side of the ball we would be actually quite certain that we won't detect label B. The key point is that in each measurement, every time we obtain some information, we 'zoom-in' into the subspace of the probability space that is consistent with so far measured values.

Continuous measurement giving continuous information. Now imagine that instead of simply moving a thumb and seeing label B, we have two types of invisible labels: non-radioactive ones A and radioactive ones B. Radioactive labels B decay with rate Γ. Similarly as before, when we just pulled one ball out of the box at time $t = 0$, everything we can tell about the ball is encapsulated in the initial probabilities $P_A(t = 0)$, $P_B(t = 0)$ and $P_{AB}(t = 0)$. But then we can surround the ball with Geiger

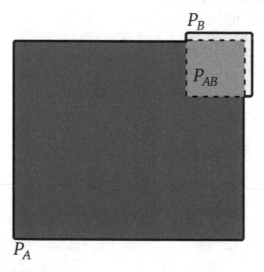

Figure 5.1. Space of possible detection events is marked with a solid line, and surfaces of red and yellow squares denoting probabilities P_A and P_B for detection of A and B events. Orange surface, overlap of the two squares marked with dashed line, corresponds to probability P_{AB} of detection of both A and B events.

counters. After a small time dt, we don't detect any clicks. This, however, gives us partial information about the system. During the time dt only a fraction of $(1 - dt\ \Gamma)$ would not decay. That means that total space describing our ball actually shrank due to the reduction of probability that the ball is in state B, changing probabilities

$$P_A(t = dt) \propto P_A(t = 0) \tag{5.1}$$

$$P_B(t = dt) \propto (1 - dt\ \Gamma)P_B(t = 0), \tag{5.2}$$

$$P_{AB}(t = dt) \propto (1 - dt\ \Gamma)P_{AB}(t = 0). \tag{5.3}$$

Now our knowledge of the system is represented with the slightly shrunken subspace (figure 5.2). To get absolute probabilities for different states of our ball, we need just to normalize different outcomes relative to the size of the new total subspace $P_A(t = dt) + P_B(t = dt) - P_{AB}(t = dt)$ (minus sign for the overlap term is because we include it twice in the simple sum of probabilities). This means that after dt time of not detecting clicks on counters, our description of the state of the picked ball evolved as

$$\begin{pmatrix} P_A \\ P_B \\ P_{AB} \end{pmatrix}_{t=dt} = \frac{1}{P_A + (1 - dt\ \Gamma)(P_B - P_{AB})} \begin{pmatrix} P_A \\ (1 - dt\ \Gamma)P_B \\ (1 - dt\ \Gamma)P_{AB} \end{pmatrix}_{t=0}.$$

Quantum case of excited atom. Let's now consider an atomic ensemble that was irradiated with a short laser pulse. We know based on dynamics that the state of the atoms is given by $|\psi\rangle = a|g\rangle + b|e\rangle$, i.e. superposition of ground $|g\rangle$ and excited $|e\rangle$ state. Atoms in excited state decay with rate Γ. We will see now how the two classical situations described above map on the case of tracking evolution of this single quantum object.

During a small time dt probability that the single atom decayed is $\Gamma dt |\langle e|\psi(t)\rangle|^2$. *If decay happened*, the state of the atom has to now be consistent with information that the excited state decayed. For the example two-level system it means that immediately after detection $|\psi(t + dt)\rangle = |g\rangle$. In other words, we zoomed into a correct probability subspace—in analogy to the classical case discussed earlier— adjusting in non-continuous manner probabilities. Because probabilities change non-continuously usually, we call this *jump*, but when discussing single quantum systems we call this *quantum jump*.

What is *quantum* in this case? Doesn't probability change abruptly even in the classical case? That's true, but there is an extra element that is quantum that we will discuss shortly in section 5.2. But before that, let's see how we account for *continuous* adjustment of probabilities in a situation when the quantum system *did not decay*. We need to have as in the classical case discussed above appropriate continuous adjustment of probability distribution that describes the system $|\psi\rangle$. With some trial and error, we can try to describe that situation with the following modification for the Schrödinger equation

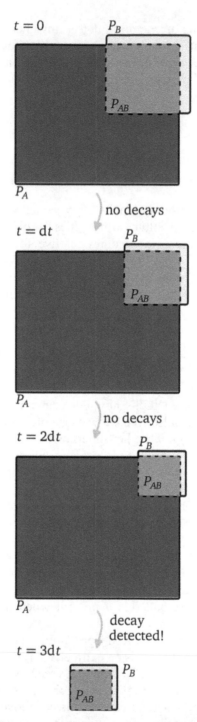

Figure 5.2. Our knowledge about the state of the classical ball selected at random from the ensemble evolving over time. The balls can have two types of labels, where A are non-decaying, and B labels are radioactive.

$$i\hbar\frac{\partial|\psi(t)\rangle}{\partial t} = \mathcal{H}_{\text{MC}}|\psi(t)\rangle, \tag{5.4}$$

$$\mathcal{H}_{\text{MC}} \equiv \mathcal{H} + i\hbar\sqrt{\Gamma}/2|e\rangle\langle e|. \tag{5.5}$$

$\mathcal{H}_{\text{MC}}$ non-Hermitian, and it doesn't maintain the norm. But it has two nice properties:

    (i) during small time step, the norm of the state shrinks by $\delta p = \Gamma dt|\langle e|\psi(t)\rangle|^2$ during the step dt relative to the norm at time t is exactly the probability to have decay event during that time; and

    (ii) in case there have been no decays, the new wave function reflects our knowledge about the system in the 'zoomed-in' subspace where no decays happened.

So taking the two together, if we want to evolve wave function at one instant $|\psi(t)\rangle$ to another instant at $t + dt$, we can write

$$|\psi(t + dt)\rangle = \frac{1}{\langle\psi(t+dt)|\psi(t+dt)\rangle}\begin{cases} |g\rangle & , \text{ decay} \\ \left(1 - \frac{i\mathcal{H}}{\hbar}dt - \frac{\sqrt{\Gamma}}{2}dt\right)|\psi(t)\rangle & , \text{ no decay} \end{cases} \tag{5.6}$$

The first factor is just normalization of the state, so that we can calculate expectations values in the usual ways. The second factor depends, as argued above, on decay happening. This methodology is useful in two main ways. Firstly, if we want to calculate density matrix—i.e. average over many realizations of identical quantum systems—we can run many simulations of the dynamics described above, simulating probabilistic decay by sampling say a random number from $[0, 1)$ and checking if the sampled number is smaller than probability $\delta p = \Gamma|\langle e|\psi(t)\rangle|^2\delta t$ that decay happened during simulation time-step δt. Only when the sampled number is smaller than δp would we choose that decay happened and adjust state appropriately. With every simulation repetition, we are doing classical averaging over many realizations of a single quantum system's state $|\psi(t)\rangle$. This is shown in figure 5.3

The second important way in which we can use the methodology above is not to calculate ensemble average, but actually really follow individual quantum systems, and calculate how their state changes, using actual decay measurements to decide how to evolve states $|\psi\rangle$ at any given time, instead of a numerical random number generator. A selection of important insights we can obtain then follows.

5.2 Effects of jumps

In this section we will focus on manifestations of new dynamics brought by the abrupt quantum jumps in strong measurements. Strong measurements project system state into a new subspace consistent with the measurement. This on its own is very much analogous to the classical case of seeing the colour of the ball drawn at random. So what is quantum here? In the classical case, probabilities describe an ensemble, whereas in the quantum case that can be a description of a

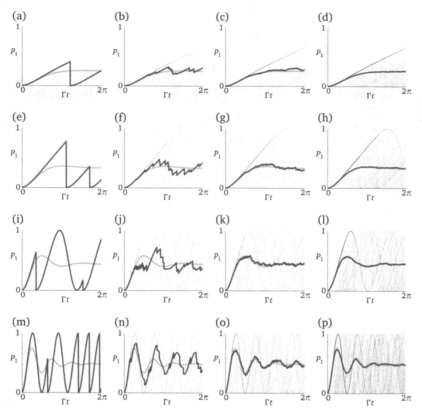

Figure 5.3. Quantum Jumps: two-level system undergoing quantum jumps. The upper level population, P_1, is plotted versus time. The rows and columns show increasing Rabi frequency and atom numbers, respectively. Rows 1–4 correspond to 1, 10, 100 and 1000 trajectories, respectively. The grey lines are the individual trajectories. The red line is the average. The blue line is the solution of the optical Bloch equations. (a)–(d) $\Omega = \Gamma/\sqrt{2}$, (e)–(h) $\Omega = \Gamma$, (i)–(l) $\Omega = 2\Gamma$, and (m)–(p) $\Omega = 4\Gamma$. An interactive figure is available at http://doi.org/10.1088/978-0-7503-2628-5.

single quantum object. But in the classical situation it is just our understanding that abruptly changes, in the case of single quantum objects, our understanding is the state of the single quantum system, and as a consequence their dynamics is going to expose these jumps in the quantum state in a way for which there is no classical equivalent.

5.2.1 Post-jump transients viewed in frequency: Mollow triplet

Let's consider maybe the simplest example of light–matter interaction: a two-level system, say atom with well defined ground $|g\rangle$ and excited $|e\rangle$ state, driven by a coherent field with Rabi frequency Ω. The excited state decays with spontaneous emission rate Γ.

Once the driving is turned on, if we observe an ensemble of atoms, we would see that atoms reach a steady state transient period of $\sim\Gamma^{-1}$. For weak driving $\Omega \ll \Gamma$ transient dynamics is smooth, reminiscent of an overdamped oscillator (figure 5.4(a)).

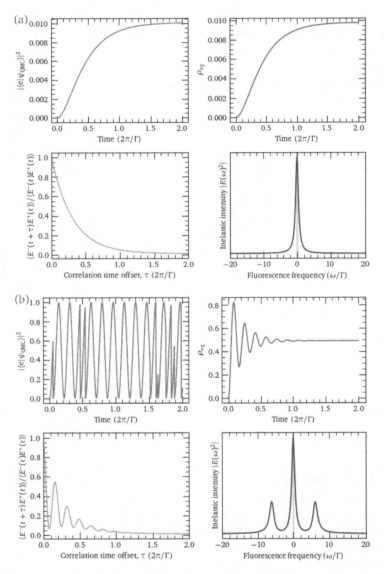

Figure 5.4. Fluorescence of two-level atom driven by monochromatic field: atom dynamics for weak (a) $\Omega = 0.1\Gamma$, $\Delta = 0\Gamma$ and strong (b) $\Omega = 6\Gamma$, $\Delta = 0\Gamma$ resonant driving. Top left panel shows an example of a single quantum trajectory. Even with no decays, the radiative damping term leads to steady state between jumps. Ensemble average of such trajectories gives a single atom density matrix (a, top right). Emitted field two-time correlation is proportional to evolution of dipole operator, and can be calculated from density matrix, and its evolution as explained in appendix A.5. Fourier transform of time autocorrelation function gives frequency spectra of fluorescence (bottom right panel). Long time correlations ($\tau \to +\infty$) give sharp elastic (coherent) component that is delta peaked at the driving frequency ω_{drive} (grey vertical line), while the remaining inelastic spectra (red line), that is not phased locked to driving field, exhibits Mollow triplet structure for strong drivings (b, bottom right). Fluorescence spectrum zero matches atomic transition. Fluorescence peak follows driving detuning for weak drivings only explore drive detuning dependence and other parameters in the interactive version avaialble at http://doi.org/10.1088/978-0-7503-2628-5.

Steady state in this case means that their coherence ρ_{eg} is not only with fixed amplitude, but also with fixed phase relative to the coherent driving field at that point: that is, atom coherence is then locked relative to the driving field phase. This is the origin of *coherent, elastic scattering*.

Directionality of coherent scattering. This is crucial in determining dominant emission direction of such an atomic ensemble. Namely, if we look over a spatially extended atomic ensemble, over several wavelengths of driving field, we will see that atoms reach steady state with a fixed phase to the atomic field at that point. But since the driving field is typically coming in the form of a propagating wave, with phase change given in space coordinate r as $\exp(ik \cdot r)$ due to driving wave vector k, the relative phase between atomic excitation will also have this relative phase. When we detect a photon, we typically cannot resolve which atom over a certain space of several $2\pi/k$ decayed. Thus emission of a photon is governed by *collective* operator $\sum_{\text{atoms}} |g\rangle\langle e| \exp(-ikr)$. When atoms are in steady state with locked relative phase to driving field, driving field imprints the phase grating that makes emission amplitude in the driving direction to always add in phase from every atom, making total amplitude $\propto N$ and emission probability in the direction $\propto N^2$. Sideways fluorescence amplitude will can be estimated as random walk of emission amplitudes of each individual atom, each contribution having a random phase and zero on average. Total amplitude will then be $\propto \sqrt{N}$, and probability of emission $\propto N$. Scattered light on the side will be $\propto \int_{\text{volume}} |\rho_{eg}|^2$, while scattered light along the driving direction will be $\propto |\int_{\text{volume}} \rho_{eg}|^2$.

Quantum jumps and transients. Now to see the signature of abrupt jumps, we just need to consider what happens *after* the fluorescence. The decayed atom starts then from ground state, and is again experiencing transient dynamics. This transient dynamics is more interesting under strong $\Omega \gtrsim \Gamma$ driving, as on average atoms will oscillate with the driving frequency Ω before their coherence reaches steady state. Keep in mind that these oscillations are happening in the rotating frame of reference, i.e. steady state means that coherence of atoms oscillates with a well defined, fixed phase relative to the laser driving field. Therefore, during the time t transient period after some emission event at time T, atomic coherence in absolute frame of reference will oscillate $\propto \exp\langle[i(t-T)\Omega\rangle]\exp(i\omega_{\text{drive}}t)$, while in steady state $\rho \propto \exp(i\omega_{\text{drive}}t)$. Since scattered light field is directly proportional to the atomic coherence, atoms in steady state will emit light matching driving frequency ω_{drive}, while atoms in transient state will emit light at atom frequency and two side peaks shifted by Rabi driving frequency $\pm\Omega$. Therefore, the fluorescence spectrum of a strongly driven two-level system will show three peaks, the so-called **Mollow triplet**, that are signatures of *transient oscillations* after each photon fluorescence event that makes the atom jump into the ground state, resetting atom dynamics. Finally, note that for a strong driving $\Omega > \Gamma$, individual quantum trajectories never reach steady state, which leads to reduction of elastic, coherent scattering, and instead of that, *inelastic scattering* dominates.

The calculation of the fluorescence spectrum (figure 5.4 (bottom right panel)) can be done based on the Fourier transform of time autocorrelation function

$|E(\omega)^2| = F[\langle E(t + \tau)E(t)\rangle]$. Since the emitted electric field is proportional to the atomic dipole operator, this can be calculated based on atom state knowledge, but importantly, is not the expectation value of several operators at the same time. Hence knowledge of the steady state density matrix is not sufficient, and one also needs to know the evolution of state and density matrix between measurements. This is a very important example of **quantum regression theorem** application for calculation of multi-time expectation values, whose details can be found in appendix A.5.

Incoherent scattering. But this is not all: what about the phase of the side peaks? Time T of any fluorescence (jump down) event is random, therefore, each atom will start these transient oscillations at random moments. Phase of this transient oscillations has nothing to do with the driving field: in other words side-bands of Mollow triplet are incoherently scattered. That is, light scattered during transient dynamics will pick random phases from each of the atoms in the ensemble that undergoes transience when we calculate decay amplitude $\sum_{\text{atoms}}|g\rangle\langle e|\exp(ikr)$ regardless of the direction of emission that we pick. Fluorescence of sidebars of the Mollow triplet is therefore distributed in a 4π sphere around atoms, in a radiation pattern characterizing a single atom σ or π transition, respectively (figure 5.5).

Comparison with classical emitters. Finally, we underline that these side peaks are a consequence of the fact that photons cannot have arbitrary small energy, but are quantized. It is this minimal emission from any given atom that projects the system in the ground state. Had the atoms been classical dipoles, charged particles on the spring driven by an oscillating electromagnetic (EM) field, no side peaks would be observed since in the classical world dynamics is continuous, without this type of extra fluctuations/transients brought by quantum jumps [2, 3].

5.2.2 Post-jump transients viewed in detection: photon anti-bunching

On an isolated single atom we can directly see transient behavior of $\rho_{eg}(T)$ after time T from start of driving/previous fluorescence, if we measure the probability of photon arrival relative to the time of the arrival of the earlier photon. This is often measured via normalized $g^{(2)}$ quantity defined such that on long time scales it goes to 1, corresponding to average photon emission $\langle n(t)\rangle_t$ rate over time t:

$$g^{(2)}(T) = \frac{\langle n(t)n(t + T)\rangle_t}{\langle n^2(t)\rangle_t}$$

Since $n(t) \propto \Gamma\rho_{ee}$ we will see that following emission detection $t \to 0$ there is no another emission of the photon before driving of the atom recovers excited state population. If driving is strong enough, we can see transient oscillations of population also in $g^{(2)}$. This is the phenomenon of photon anti-bunching following fluorescence from a single two-level system, and is another consequence of the abrupt state transition (in this case to ground state) following strong measurement performed by detection of a fluorescence photon [4].

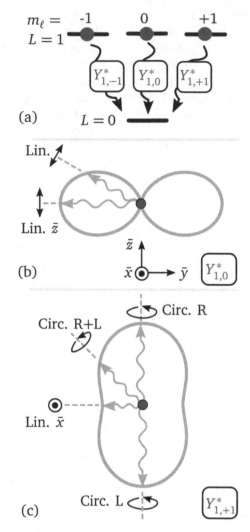

Figure 5.5. Spatial fluorescence patterns from single-atom decays in the far-field. (a) Level scheme of an excited atom decaying under dipolar coupling $\propto \langle L = 0, m_\ell = 0 | Y^*_{1,q} | L = 1, m_\ell \rangle$, $q = 0, \pm 1$ into the vacuum modes. Solid line (yellow) on (b) and (c) are polar plots of fluorescence *intensity* in $\bar{y} - \bar{z}$ plane for $q = 0$ and $q = +1$ decays, respectively. The scale is the same on both plots, and the fluorescence pattern is $\bar{z}$-axially symmetric. Polarization of emitted light in several example directions (wavy lines) is labeled Lin., Circ. R, Circ. L, corresponding to linear, right-hand circularly polarized and left-hand circularly polarized, respectively.

5.2.3 Jumps communicating information overlap: entanglement and interference

So far we considered jumps in the system state brought by strong measurements that were conveying unequivocal information: bang! we detected a photon which means that this atom decayed to ground state. However, sometimes many different events can be overlapped in one measurement channel: for example we might have two excited atoms in our experiment, and a single detector facing both of them. Now detection of a single photon tells us that either the first atom OR the second atom decayed. If we accept statement in the introduction, that 'things happen in the world consistently with available information about them', it is understandable that after the measurement system, the state of the two atoms will jump into the state where one or the other thing happened before, in other words atoms will become entangled. We will consider now several examples where we deliberately engineer or naturally have overlapped communication channels, and consequences of jumps.

Entanglement by measurement. Consider a glass cell filled with alkali atom vapour (figure 5.6(a)). We can consider a subset of internal atomic states that gives a so-called Λ scheme (figure 5.6(b)), consisting of two hyperfine ground states $|a\rangle$ and $|b\rangle$, and one excited state $|e\rangle$. Transitions between the hyperfine ground states correspond to the microwave regime, and typically happen because of the atom–surface collisions. The collision rate can be suppressed by adding some buffer gas which increases the time it takes atoms to diffuse towards the glass wall. This makes hyperfine ground states long-lived. These states can be used in a form of atomic memory in the following way: the whole atomic population in the vapour cells can be prepared, for example by optical pumping, into one of the two ground states, for example $|a\rangle$. We can store stochastically a single excitation by irradiating the vapour cell with an off-resonant laser and waiting until one photon is detected (figure 5.6(a)) signaling that one atom, somewhere in the atomic medium along the detection direction, is transferred into an excited state $|b\rangle$ (figure 5.6(b)).

Now, because in the optical setup presented in figure 5.6(a) emission from any atom in the direction of the detector will result in detector click, we don't have information which exact atom decayed. In this case, jump signals action of *collective* operator: $\sum_{\text{atoms } j} |b_j\rangle\langle a_j| \exp\left[i(k - k_d)r_j\right]$ where r_j is the position of jth atom, and k is a emission photon wave vector directed into the detector, and k_d is the driving laser wave vector.

To release a stored photon from this atomic memory, we would switch off the initial driving field, and add a readout driving field with wave vector k_r, chosen such that it is near-resonant with $|b\rangle \to |e\rangle$ transition. The key question now is *what sets direction of the readout photon wave vector k'?* To calculate the amplitude of the readout in direction, since we don't resolve again atom emission within ensemble, we would act with a collective operator $\sum_{\text{atoms } j} |a_j\rangle\langle b_j| \exp\left[i(k' - k_r)r_j\right]$. At the end (figure 5.6(d)) all atoms are again in the initial state $|a\rangle$. In other words, there is no information left about which atom was excited *not only in our measurement apparatus*, but also *anywhere else in the physical system*. This is important, because then the final probability amplitude is then summed (interference) over all small probability amplitudes corresponding to the situations when the jth atom was excited, each one contributing with relative phase of $\exp\left[i(k - k_d + k' - k_r)r_j\right]$. Because atoms are randomly distributed in space r_j, the only way for the final probability amplitude to be large is if for some direction of k' the sum of the wave vectors is zero $k - k_d + k' - k_r \to 0$.

To summarize the main points of the argument above, it is interference between many possible excitation paths that causes that ultimately one direction in space is favored for the readout. Before the readout the atoms in the medium are in an *entangled* state, with well defined phases *relative* to their initial phases. Note that

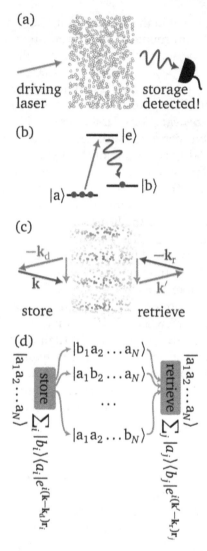

Figure 5.6. Atom vapour quantum memory. Driving laser illuminates an ensemble of atoms whose fluorescence decay—signaling storage of a photon—can be detected with a single-photon counter (a). Initially all atoms are in $|a\rangle$ state, and single fluorescence photon detection signals storage of one excitation in the ensemble (b). Mismatch between drive and store field wave vectors imprints phase grating on relative phases of entangled atoms collectively storing single excitation (c). The whole atomic ensemble acts as a giant many-path interferometer, which ultimately produces constructive interference only for one retrieved field direction k' (d).

also the argument does not depend on initial phase of probability amplitude of any atom's initial state $|a_j\rangle$: the only thing that mattered was the relative phase in the 'interferometer' in figure 5.6(d), not the input and output phase. We will see later that what matters are always these relative phases—between different paths—that accumulate when system evolution splits into different paths, until these paths are recombined again later, and that often the initial starting phase is not defined. It was important that input and output state in this interferometer are the same, so that every atom contributes coherently (as probability amplitude sum) in the final amplitude for emission.

More generally, this is an example where we create entanglement in the system through the act of measurement that does not resolve different options. Usually, one way to create entanglement between parts of the system is to let two systems interact. In this case individual atoms *were not directly interacting*. Instead we brought these N systems—if each atom is considered to be labeled as a system—into an entangled state by letting them all interact via the third shared system (same radiation field mode). After projective measurement brought by detection of a scattered photon that heralds storage of excitation in the atomic medium, the collective system state jumped abruptly into the entangled state.

Note that this type of entanglement discussed above is actually quite common. For example, when learning about lasers, it is often taken in simplified treatments that stimulated fields have phase, direction and frequency equal to the incident field. Of course, a well defined beam of light cannot be produced by a single atom as any dipole transition pattern is not nearly as selective as a directed laser beam. In *Laser Physics* [5] Sargent, Scully and Lamb highlight that one should discuss not single atoms but 'chunks' of pumped media: 'these three properties—equality of phase, direction, and frequency of incident and stimulated fields—are completely obtained only for appropriately pumped media whose cross-sectional areas contain many similar dipoles'. There too we obtain directionality on the basis of the well defined relative phases for excitation among different entangled atoms[4].

Finally, we note how this principle of measuring information overlap can be extended. In the above example, overlap in measurement results was a natural consequence that we had a lot of atoms emitting in the same direction, without capability on the measurement optics side to resolve individual emissions. Note that what we use on the measurement side to guide and overlap information is important and can be used to establish different entanglements. For example, imagine we have two identical vapour cells as above, both illuminated with a driving laser that tries to store excitation in the respective medium (figure (5.7)). Imagine that now for detecting a scattered photon that heralds the storage we use one photodetector for both of them. For example, by combining the optical signal from both cells on a beam splitter that overlaps perfectly two geometric modes of the propagating herald field. In that case, after detecting a single photon, we don't know which of the two memories stores excitation. And indeed the correct state is the entangled state where we have superposition of two possible outcomes, where one of the two memories stored the photon[5].

---

[4] Another reason why entangled states of the type $\sum_i |a_0...a_{i-1}b_i a_{i+1}...\rangle$, also called **W-state**, are common is because measuring n atoms in this state most likely result in measurement outcome $|a\rangle$, and that measurement just projects the big state to the sub-state where now the remaining $N - n$ atoms are in the W state. This is in stark contrast to so-called maximally entangled states, which would completely collapse to a single state by measuring only a single component of the system.

[5] Schemes like this are explored for distributing quantum entanglement across large distances in the field of quantum networks. See, e.g. [6, 7].

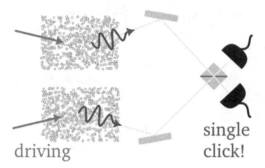

driving   single
                                                     click!

Figure 5.7. DLCZ protocol for photon storage of a single photon in two entangled atomic memories.

If in doubt, interfere! (In)distinguishability in space, time and frequency. In the previous section we looked into an interference pattern in space, and emphasized that we were adding amplitudes, instead of just summing probabilities, because we got an identical atomic state $\otimes_{j=1...N}|a_j\rangle$ regardless of which atom decayed, leaving no information about emission direction in the atomic medium. Here we want to look at two other common examples of interference, and to explore situations in which some information stays in the atomic medium.

Consider two different types of atoms, both types having three internal energy levels, but one organized in a V-scheme and the other in a Λ-scheme (figure 5.8). Imagine that atoms are initially in state $|a\rangle$ and we drive both atoms with a very short laser pulse, so short that it has bandwidth sufficient to address both excited states in the V-scheme and populate them. We then start the timer and wait for decay to happen, accumulating data for a histogram of photon arrival times over many identical repetitions of the atom state preparation with short driving pulse. What would the distribution of such photon arrival times look like?

To start, let's see what the classical picture gives us. In both cases there are two transitions that can happen, with corresponding dipole moments. If we imagine two classical dipoles, identically prepared every time, radiating while their oscillation energies differ by Δ, we would notice beats in emission of radiation. Namely, in addition to fast, optical frequency oscillations, amplitude of oscillations would be slowly changing with frequency Δ corresponding to frequency difference between two classical oscilators[6]. In the classical picture, in both situations we would see beats (oscillations) in the photon arrival time histogram, with the period of $2\pi/\Delta$.

Quantum prediction for the two cases, however, differs because of the key difference that now each atom can decay only once, and upon decay it may or may not leave information about which transition occurred. In the V-scheme, regardless of initial state, the atom ends up in state $|a\rangle$. With no information about which transition happened stored anywhere, the final outcome will be interference between probability amplitude for decay from $|b\rangle$ and $|c\rangle$ state, leading to beats in photon arrival times, as predicted also in the classical case. However, in the Λ scheme, after

---

[6] This is a consequence of the simple relation $\cos(\omega_1 t) + \cos(\omega_2/t) = 2\cos\left(\frac{\omega_1 - \omega_2}{2}t\right)\cos\left(\frac{\omega_1 + \omega_2}{2}t\right)$.

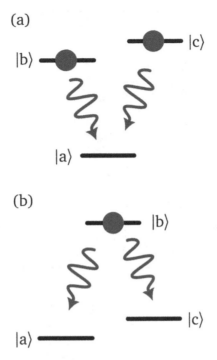

Figure 5.8. Distinguishable and indistinguishable paths. For a single atom prepared with a short pulse in the superposition state of two excited states in V-level scheme (a) one cannot distinguish between two decay paths ending in the same state. Single atom quantum beats occur then. On the other hand, in Λ scheme (b) which-path information is stored in final atomic state |a⟩ or |a⟩, hence no single-atom beats are detected in fluorescence, binned based on time of detection after excitation pulse.

each atom decay one could in principle go and check if the atom ended up in state |a⟩ or in state |c⟩. Thus, there is no interference between these two probability amplitudes because of which part of the information is now stored. Instead we will see superimposed two exponential decays, corresponding to decay along one of the two routes. Each time route is selected randomly, but there won't be interference between them. Importantly, this is the case *regardless* of us looking up or not in what state an atom ends up after emission. In other words, even if which-path information is not recorded by us, but is recorded somewhere in the system degrees of freedom, dynamics is unfolding consistently with total (in principle) accessible information.

Finally, let's consider another way to detect that situation when a quantum system is in a superposition state, that works without the need for pulsed excitation and time resolved detection: steady state absorption spectra. Consider for example having an ensemble of atoms whose internal structure is given by the V-scheme, and whose states |b⟩ and |c⟩ have very different lifetimes, such that |b⟩ is long-lived, and |c⟩ whose decay rate Γ is large, that its, spectral line width $\sim\Gamma$ spreads over more than Δ that is separation between the two excited energy levels. If we look at the steady state weak-probe absorption spectrum, we will see a narrow and a broad line shape that *are not* simply two line shapes superimposed, but typically something more asymmetric. Because both possible transitions are driven with the same laser, they

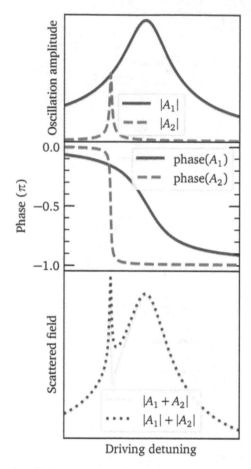

Figure 5.9. Fano resonance. Amplitudes A_1 and A_2 of two driven harmonic oscillators (top panel), showing characteristic π phase shift (middle panel) relative to driving when driving crosses resonances. For charged oscillators amplitudes are proportional to scattered fields. If oscillators' relative phase is not fixed, the scattered field is proportional to the sum of the absolute amplitudes (dotted, bottom panel). If their relative phase is fixed, the scattered field exhibits interference, and the characteristic shape is called Fano profile (full line, bottom panel).

have well defined relative phase, but in the quantum case, just as in classical case, when the drive field sweeps through the resonance of the oscillator, the phase of the oscillator changes by π. The forward field after the ensemble of the atoms, that is the sum of the driving field and two scattered fields from the two possible atomic transitions. As narrow transition resonance amplitude changes phase as driving sweeps through the first resonance, it will interfere with the field emitted from the second transition, that being further down from the resonance changes much more slowly and acts as a phase reference in this measurement. That gives rise to the specific profile shown in figure 5.9 that is often called **Fano resonance** (named after Ugo Fano (1912–2001)). Fano resonance is in general a wave phenomenon, and works also when there are two classical oscillators that simply have well defined relative phase obtained from a common driving field. So while Fano resonance does not imply necessarily quantum correlations, it is a signature that two resonances have some common cause, providing a well defined relative phase, which can be useful when deducing microscopic dynamics.

Phase from collective decays: superradiance. So far we discussed the consequences of one or two jumps on dynamics. Now we will consider what happens when there is a sequence of jumps. Consider an ensemble of N two-level atoms with ground $|g\rangle$ and excited $|e\rangle$ state. Atoms are separated randomly in large space, so that they are many wavelengths of $|e\rangle|g\rangle$ transition away. Under these conditions, an atom doesn't feel directly the fields from other atoms, and each of their excited states decays with spontaneous emission rate Γ determined for single isolated atoms.

What happens if we prepare all atoms, for example by short laser pulse, in the excited state and let them evolve? Individual atom emissions are not determined in space beyond the standard radiation pattern of an isolated dipole. But once one decay, at random, happens, something new starts to play in the dynamics: entanglement between the different atoms brought by collective decays. Consider for example the state after n consecutive emissions in a given direction k. If initial spontaneous radiation rate *in that direction* was NI_0—where I_0 is single-atom radiation rate—after n decays in that direction, the spontaneous rate will be $n(N - n)I_0$. The emission rate in the chosen direction *increases* (figure 5.10(b)).

This is clearly seen if one organizes an N-atom basis in the so-called **Dicke ladder**[7]. Dicke noticed that we can organize 2^N state basis of the N two-level systems decaying radiatively through collective decays in the following manner. By analogy of N spin-1/2 systems, we can introduce operators analogous to angular momentum operators L^2, L_x, L_y, L_z that are defined by:

$$R_{x,i}|\ldots(e_i/g_i)\ldots\rangle \equiv \frac{1}{2}|\ldots(g_i/e_i)\ldots\rangle,$$

$$R_{y,i}|\ldots(e_i/g_i)\ldots\rangle \equiv \pm\frac{1}{2}i|\ldots(g_i/e_i)\ldots\rangle,$$

$$R_{z,i}|\ldots(e_i/g_i)\ldots\rangle \equiv \pm\frac{1}{2}|\ldots(e_i/g_i)\ldots\rangle.$$

Similarly, total $R^2 \equiv \sum_i(R_{x,i}^2 + R_{y,i}^2 + R_{z,i}^2)$, and collective raising and lowering operators $R_\pm \equiv R_x \pm iR_y$. We can organize all 2^N states based on values of R^2 and m_z value of $\langle L_z \rangle$ (figure 5.10). Starting from a completely inverted state $|e_1\ldots e_N\rangle$ collective decay action is proportional to the action of the lowering operator L_-, and selects N states that have the same value of R^2. In the $m_z = N - 1$ manifold there are remaining $N - 1$ states[8] with $R^2 = N/2 - 1$. In this Dicke ladder basis, we can clearly see how the sequence of collective decays leads to states that have decay rates

---

[7] Introduced by Dicke (1916–97) [8].
[8] Except for the maximally symmetric manifold, (R^2, m_z) does not uniquely identify states, thus some manifolds are degenerate.

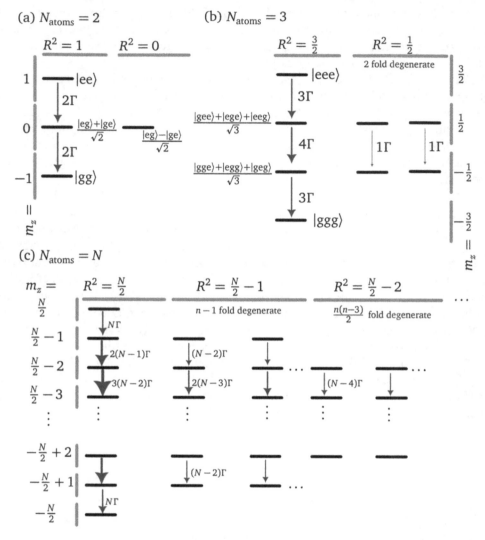

Figure 5.10. Dicke ladder, superradiance and sub-radiance. Example Dicke ladder for two atoms. $R^2 = 1$ corresponds to symmetric collective states decay, but we notice also antisymmetric state ($R^2 = 0$, $m_z = 0$) whose decay amplitudes destructively interfere—this state does not radiate at all. Example Dicke ladder basis for three atoms (c) shows how after initial emission of symmetric inverted state the decay rate is increased to 4Γ, which is higher than initial rate 3Γ of three excited atoms due to constructive interference between decay amplitudes. In $R^2 = 1/2$ manifold we have orthogonal states with also two excited atoms, but that have decay rates of 1Γ, lower than expected rate of 2Γ for two excited atoms. These are subradiant states. This basis can be generalized for N atoms, where build-up of stronger decay rates between states is indicated with increased thickness of lines indicating decay rates, clearly showing a superradiant cascade. This leads to build-up of so-called superradiant flash whose peak decay rate is equal to $\frac{N}{2}(\frac{N}{2} + 1)\Gamma$ for even N.

not only stronger than non-entangled N-atom ensembles with corresponding number of excited atoms, but also stronger decay rates than the initially inverted ensemble. These are superradiant collective states. We also see that some states are more weakly decaying than corresponding non-entangled states—these are

subraddiant states. Indeed, disregarding spatial positioning of atoms for the moment, for odd atom numbers there are always states whose possible decays to other states cancel completely—these are non-radiative collective states.

Note that electric field expectation value $\langle E \rangle$ for decay of any of these states is zero, but decay intensity, or photon-count $\propto \langle E^\dagger E \rangle$ is non-zero. This is because states that have a fixed well defined number of excited atoms have zero dipole matrix expectation value since their phase is completely undetermined. So while semi-classically such states wouldn't decay, in quantum treatment and experiments they do.

The above considerations strictly hold true for emitters closer than λ that don't interact. For full treatment one has either to account for atom interactions, or consider emitters that are far apart and non-interacting. In spatially extended ensembles, few initial emissions will be at random, since there are many potential decay modes. However, directions of their initial collective decays start breaking symmetry on the system and imprinting a phase pattern (section 5.2.3) that leads to the increased probability that the next emission will be in that same direction. As a consequence, symmetry gets broken and a preferred emission direction is established. If we measure light intensity in that direction, we will not see just the usual exponential decay, which corresponds to individual atom deexcitation. Although for extended samples where atom–atom interactions are negligible, every single atom still decays with a fixed rate Γ, the relative phase imprinted on atoms by consecutive collective decays reduces emission probability in directions other than the one selected by the broken symmetry. As a consequence, in that direction we see the initial start of exponential decay in emission, followed by a bump marking increased emission in that direction, before it again returns to exponential decay at longer time scales. This phenomenon is called **superradiance** and the bump is called **superradiant flash**. In experiments symmetry is often broken deliberately to pick the preferred direction for initial emissions, either by using clouds of emitters elongated along some specific axis, or by adding some initial weak 'seeding' laser.

We see that consequences of consecutive *collective* spontaneous decays can be quite dramatic: leading to spontaneous symmetry breaking and qualitatively new features in dynamics.

Information communication via non-orthogonal basis states. The idea of information measurement communicating information overlap—conjugate coding, introduced by [9]—is in a crucial way used in protocols for **distribution of encryption keys via quantum protocols** such as BB84[9]. The protocol allows for distribution of a shared secret—some random binary sequence—between two parties that want to establish safe communication. A random sequence generated by one of the parties is transmitted to the other party encoded into polarization degrees of freedom of single photons: two orthogonal directions indicating 1 and 0. Now crucially, the first party can send this information encoded in two different basis sets, offset by 45° (figure 5.11). The receiving party also at random decides the measurement basis. In the case where the measurement and encoding basis match, received polarization

---

[9] Named after initials of the authors, Charles H Bennett and Gilles Brassard, who devised the protocol in 1984.

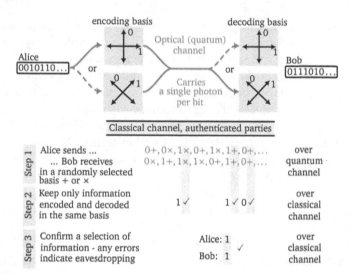

encoding basis Optical (quatum) channel decoding basis

Alice 0010110...   or   Carries a single photon per bit   or   Bob 0111010...

Step 1	Alice sends ... ... Bob receives in a randomly selected basis + or ×	0+,0×,1×,0+,1×,1+,0+,... 0×,1+,1×,1×,0+,1+,0+,...		over quantum channel
Step 2	Keep only information encoded and decoded in the same basis	1✓	1✓ 0✓	over classical channel
Step 3	Confirm a selection of information - any errors indicate eavesdropping	Alice: 1 Bob: 1 ✓		over classical channel

Figure 5.11. BB84 secret key distribution protocol. Secrecy relies on sending information encoded in single photons, and no-cloning theorem for quantum states. Anyone who intercepts communication on quantum channel will not be able to copy and measure state of the photon without projecting the original photon.

—and correspondingly encoded bit—will be the same as the one that was sent. However, if the measurement and encoding basis are offset by 45°, then the photon sent in either of the initial encoding polarization states has equal probability of ending as 0 and 1 in the measurement basis. That is, the two possible information channels are perfectly overlapped on both measurement channels, which completely scrambles initial sent information, and produces a random measurement sequence.

The BB84 protocol starts with sending single photons that encode a randomly generated sequence in initially randomly selected basis sets, measured in another randomly chosen measurement basis. After the measurement two parties communicate, over a public channel, selected encoding and measurement polarization basis, and keep only bits that were sent and received in the matching basis. To ensure that there was no eavesdropping attack on their single photon stream, they also communicate randomly selected selection of the measurement sequence for which the chosen basis match. Any eavesdropper, trying to measure photons in between two parties, and producing a new photon for the receiving party, could not reproduce the original photon polarization based on that single measurement (no-cloning theorem) and would therefore introduce errors visible in part of the publicly communicated subsection of the sequence. If such errors are not present, two communicating parties have now a shared random sequence that they can use for standard, classical encryption protocols.

Scrambling of information is also used in the **quantum teleportation** protocols. Here state $|\psi\rangle_C = \alpha|0\rangle_C + \beta|1\rangle_C$ is mapped to another qubit without two qubits directly interacting with each other. Two states can be for example encoded in two orthogonal polarizations of a photon. The initial resource for this is an entangled pair of qubits A and B that is initially prepared in one of the entangled Bell's states, and that are then distributed on two locations between which teleportation is to happen. Measurement of particle A with C in the Bell basis completely scrambles the initial state of particle C, and leaves two classical bits of information. Simultaneously, due to entanglement, the state of particle B will be projected into one of the four options depending on the outcome of the initial Bell measurement. One of these states is identical to state C while the three others can be simply converted, with single qubit gates, to state C. Knowing two bits of information obtained in initial Bell basis measurement is sufficient to know what to do on location B to recover state A.

To see this, we will make use of two-spin Bell state basis

$$|\Phi^+\rangle_{AB} = (|0_A 0_B\rangle + |1_A 1_B\rangle)/\sqrt{2} \qquad (5.7)$$

$$|\Phi^-\rangle_{AB} = (|0_A 0_B\rangle - |1_A 1_B\rangle)/\sqrt{2} \qquad (5.8)$$

$$|\Psi^+\rangle_{AB} = (|0_A 1_B\rangle + |1_A 0_B\rangle)/\sqrt{2} \qquad (5.9)$$

$$|\Psi^-\rangle_{AB} = (|0_A 1_B\rangle - |1_A 0_B\rangle)/\sqrt{2} \qquad (5.10)$$

Let's assume that an entangled pair of photons A and B—distributed in advance between two locations that are participating in teleportation—is in state $|\Phi^+\rangle_{AB}$. Total state of the system of three photons A, B and C can be written then as

$$
\begin{aligned}
|\Phi^+\rangle_{AB} \otimes |\phi\rangle_C = &\ |\Phi^+\rangle_{CA} \otimes (\alpha|0_B\rangle + \beta|1_B\rangle) \\
&+ |\Phi^-\rangle_{CA} \otimes (\alpha|0_B\rangle - \beta|1_B\rangle) \\
&+ |\Psi^+\rangle_{CA} \otimes (\alpha|1_B\rangle + \beta|0_B\rangle) \\
&+ |\Psi^-\rangle_{CA} \otimes (\alpha|1_B\rangle - \beta|0_B\rangle)
\end{aligned}
\qquad (5.11)
$$

We see that amplitudes in the second part of the product in all four terms are very similar to the initial state. In fact if we perform Bell state discrimination measurement on a pair of photons A, C in one location, and detect one of the four Bell states —'scrambling' information on that end—we will project the state of the photon B in another location in one of the four similar states. If we send (classical) information about which of the four possible Bell states are obtained on the first location, then we can apply feedback on a single photon B to map its state to the initial state of the photon C. We would perform nothing (identity), Z, X or Y single qubit gate on photon B for $|\Phi^+\rangle_{CA}$, $|\Phi^-\rangle_{CA}$, $|\Psi^+\rangle_{CA}$ or $|\Psi^+\rangle_{CA}$, respectively. That finishes mapping, or teleportation, of information from photon C to photon B.

5.2.4 Quantum non-demolition measurements

The measurements so far typically obtained information about a photon at the cost of detecting that same photon. There is also another type of measurement where relevant information is mapped, through some form of interaction, to another system, and then another system is being measured. For example, one optical field can induce phase shift—on another field nearby—that will be proportional to the intensity of the field, thanks to the Kerr nonlinearity existing in various materials. If we use some interferometer setup then to measure phase shift of the second probe field, we can determine intensity of the initial field without absorbing a single photon from that field.

Some of the most precise measurements like this have been demonstrated using highly excited, Rydberg states of the atoms[10], in the microwave fields of only a few photons. Rydberg atoms have been prepared in superposition of two highly excited, Rydberg states, and then they would be sent through the microwave cavity that contains a certain number of microwave photons. Some transitions between Rydberg states are in the microwave range of the EM spectrum. However, since we don't want to absorb the microwave photons in the cavity, since that would destroy them, we can choose Rydberg states such that they don't have a resonantly driven transition available for the specific energy of microwave photons in the cavity. Yet, these microwave photons will still couple to the atoms in the second order of interaction, via off-resonant states, giving rise to slight energy offset of the Rydberg state energy, proportional to the intensity, i.e. number, of microwave field in the cavity. If the two chosen initial Rydberg states of an atom have different constants of proportionality, the relative phase offset between the two states can be measured to determine the number of photons in the cavity.

Measurement occurs one atom at a time, and every time only a tiny part of the information is extracted from the system. We initially apply $\pi/2$ pulse to prepare the system in the superposition of two Rydberg states $(|r_1\rangle + |r_2\rangle)/\sqrt{2}$. If we let an atom with velocity v pass through the cavity of length L, if the cavity contains n photons the result state will be $|\psi\rangle = \exp(i\alpha tn)|r_1\rangle + \exp(i\beta tn)|r_2\rangle$, where α and β are constants, depending on the Rydberg state, that characterize phase shift that states accumulate over time t. After the atom leaves the cavity at time $t = L/v$, we apply another $\pi/2$ pulse—closing in this way the interferometer loop we created before in the internals states of an atom (figure 5.12)—and measure probability that the atom is in state $|\langle\psi|r_2\rangle|^2 \propto \cos^2\left[(\alpha - \beta)\frac{L}{v}n\right]$.

Interestingly, detection of a single atom in excited state $|r_1\rangle$ or $|r_2\rangle$ does not give us immediately the above-mentioned probability, and hence we do not immediately find the value of n. By using the quantum Monte Carlo trajectory approach (see beginning of this chapter, section 5.1), we can calculate exactly how our understanding of the cavity photon number evolves after every detection. We can either use results of the experiment, to reconstruct cavity photon state over time, or use a random number generator to reconstruct one possible trajectory. We do the second thing, using our numerical experiment to track evolution of the state of the cavity with every consecutive single atom probe. The state of the atom–field system after the atom has passed through the cavity and second $\pi/2$ pulse, but before its state is measured, has the entangled form of

---

[10] For more about Rydberg states see e.g. [10].

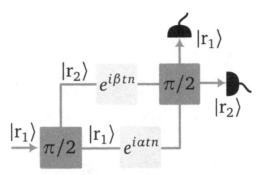

Figure 5.12. Constructing two-path interferometer using internal states of Rydberg atoms that experience different phase shifts due to n off-resonant microwave photons in the cavity. Atoms actually travel through the same space, but the overall state follows two paths due to differences in internal states.

$$|\psi\rangle = \sum_{n}\left\{\cos\left[(\alpha - \beta)\frac{L}{v}n\right]\phi(n)|n \otimes r_2\rangle + \sin\left[(\alpha - \beta)\frac{L}{v}n\right]\phi(n)|n \otimes r_1\rangle\right\}$$

where $\phi(n)$ is the probability amplitude that the number of photons in the cavity is n. We see in figure 5.13 how after every measurement, there is a small renormalization happening as state of the system becomes consistent with the observations so far— that renormalization is called **quantum back action**. Gradually, the cavity photon state is evolving scholastically towards one of the stable number states. States like this, that *for specific probing interaction and measurement scheme* are resistant to measurements, are also called **pointer states**. This terminology is used even when we deal with standard dissipative measurement schemes, where the system is measured by the environment through spontaneous decay for example. Engineering dissipative interaction can steer a quantum system to different states, and it represents one possible way of stochastic **quantum state preparation**.

5.3 Effects of information leakage: new coherent dynamics

In section 5.2 we explored some effects arising from 'step-like' jumps in system evolution, brought by abrupt changes in our knowledge of the system through strong measurements like that of detection of a fluorescence photon.

In this subsection we will completely focus on what happens to coherent dynamics in the case when no such strong measurements are made. In the absence of observation we still get some information about the system, but at a slower pace. If we make an analogy, imagine you want to determine the number of dancers in a dance hall. You might take a photograph of the dance hall from above and immediately determine that there are three couples dancing. This is what happens in strong measurements.

In an alternative approach, we are peeking through some keyhole and seeing just part of the room. At any given moment we don't see a single dancer in that part of the room, but we know that dancers over time move through the room, so if we wait for a long time we should be able to see all the dancers. Equally, for every moment

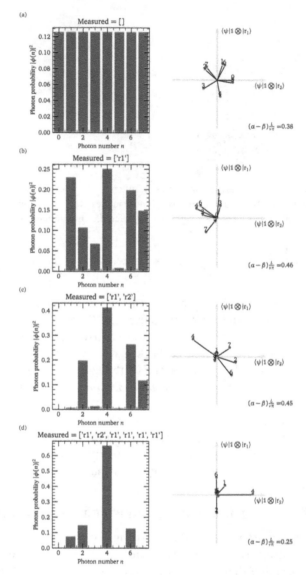

Figure 5.13. Quantum non-demolition measurement of photons in cavity and quantum back action. Starting from initially equally populated state of the photon modes $\phi(n) = \frac{1}{\sqrt{7}}$, individual atoms in superposition of two Rydberg internal states $|r_1\rangle$ and $|r_2\rangle$ are interacting for varying time, depending on their velocity v, with a field in the cavity. An example trajectory is shown in panels (a and b). Phase shift per photon for the current probe atom is indicated in the right panel (bottom right corner), while initial photon probability before measurement is shown in the left panel. Just before the measurement, atom–cavity field state function $|\psi\rangle$ consists of a number of different amplitude contributions for different photon numbers n in the cavity, which will contribute different projections to final detection amplitudes of probe atoms in one of the two Rydberg states. After measurement, only projections of amplitudes on the measurement axis corresponding to the obtained measured atom state survive, and are renormalized. After a number of measurements, the state photon number in the cavity is well defined (d). An interactive figure is available at http://doi.org/10.1088/978-0-7503-2628-5.

that passes without seeing dancers, we are becoming more and more certain that there is no one dancing there. Now what happens in quantum measurements under no detection is that we are receiving similarly partial information that modifies probability amplitudes: making outcomes we don't see less likely for example, and, importantly, *renormalizing* the resulting state. Thus continuous observation in the absence of abrupt new information of the system still continuously shapes coherent dynamics. Here there won't be jumps associated with abrupt changes in our knowledge of the system. But coherent dynamics can be affected drastically as we will see in the following examples.

One important point of equation (5.6) introduced above is that if we have a very good experiment we can really calculate in this way the wave function of the system that evolved under no-clicks. This became relevant for experiments only recently. Usually, in optical systems, it is hard to capture all possible decays since they are not very directional (covering typically 4π steradians, usually in some of the typical patterns for dipolar emitters) and there are some losses in the optical detection chain. This precludes certainty in knowing if the single quantum system underwent optical transition during some time in the experiment. However, for solid state qubits interacting with microwave fields, as in superconducting circuits, one can capture all decays with much higher fidelity.

Consider a three-level system in so-called V-configuration (figure 5.14): with one ground state $|g\rangle$ and two excited states $|e\rangle$ and $|r\rangle$. The coherent driving with Rabi frequency Ω_D couples $|g\rangle \leftrightarrow |e\rangle$ transition. To measure the state of the system we introduce coupling to the readout state $|r\rangle$ that radiatively decays with constant Γ into some waveguide from which we can detect the atomic state. The basic idea is that if the system is in ground state $|g\rangle$ then adding an extra driving on $|g\rangle \leftrightarrow |r\rangle$ transition, will allow us to detect that via spontaneous emission of state $|r\rangle$.

The interesting point is how the coherent dynamics—when no spontaneous decays are detected—changes just due to the fact that we are performing this measurement through readout state $|r\rangle$, as a consequence of the renormalization of the state upon no-click measurement. In the case where readout driving is turned off, or is very weak, $\Omega_R/\Omega_D \to 0$, $\Omega_R/\Gamma \to 0$, we have standard coherent **Rabi oscillations** between two states with angular frequency Ω_D (figure 5.15(a)). Now, if we start with a system in ground state, and have strong readout $\Omega_D < \Omega_R$, $\Omega_R/\Gamma \ll 1$, we experience **speed-up of coherent dynamics**: *when no clicks are detected the system*

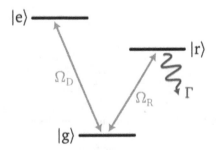

Figure 5.14. Three level system with V-configuration of energy levels.

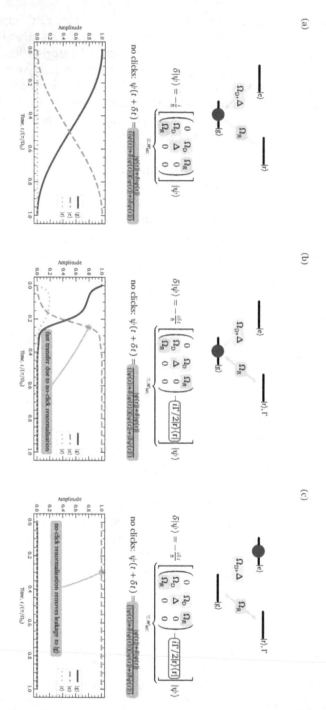

Figure 5.15. Coherent (no-click) dynamics. Top panels show energy level diagram and initial state population. Middle panels show evolution dynamics equations, and the bottom panels show a single wave function trajectory in the case where no decays (no clicks) have been detected from the system. (a) Under $\Omega_R = 0$, $\Gamma = 0$, $\Delta = 0$ and initial state $|g\rangle$ we observe standard Rabi oscillations. (b) Strong coupling to

readout state $\Omega_R = 20\Omega_D$, $\Gamma = 50\Omega_D$, $\Delta = \Omega_R/2$, with initial state $|g\rangle$, display speed-up of transfer to $|e\rangle$ state *when no decays are detected*, in spite of fixed Ω_D. This is solely due to renormalization of the state under no-click dynamics. (c) If we keep strong coupling to readout state $\Omega_R = 20\Omega_D$, $\Gamma = 50\Omega_D$, $\Delta = \Omega_R/2$, but start from state $|e\rangle$, the population will be stuck in the excited state $|e\rangle$ since dynamics is reset essentially in every time step due to no-click dynamics that renormalizes wave function such that the population is in state $|e\rangle$). An interactive figure is available at http://doi.org/10.1088/978-0-7503-2628-5.

goes to the excited state coherently much faster than would be expected on just driving rate Ω_D: this is the effect of renormalization of the state on the subspace that is consistent with no-clicks outcome, which coherently zooms the possible system space into a case where being in the ground state is increasingly unlikely (figure 5.15(b)). This has been recently verified experimentally [11].

Finally, if measurements are very strong, $\Omega_D \ll \Omega_R$, $\Omega_R/\Gamma \ll 1$, and we start initially in the excited state $|g\rangle$, the state renormalization due to observed no clicks almost completely freezes dynamics (figure 5.15(c)). Effectively, for every time step, if the system was in the ground state, we would with high probability detect spontaneous decay of the readout state. In a situation without clicks, it means that state space consistent with no-clicks dynamics has a projection that essentially has no contribution of the ground state. This resets the transfer of population under Ω_D during the corresponding time step. This phenomenon is known as **quantum Zeno effect**.

5.4 On phase in quantum physics

> *...let's imagine that we have a stopwatch that can time a photon as it moves. This imaginary stopwatch has a single hand that turns around very, very rapidly. When a photon leaves the source, we start the stopwatch. As long as the photon moves, the stopwatch hand turns (about 36 000 times per inch for red light); when the photon ends up at the photomultiplier we stop the watch. The hand ends up pointing in a certain direction. That is the direction we will draw the arrow.*
>
> Richard Feynman explaining calculation of contributions for probability amplitude for a given path, [12] p 27.

The phase, or more precisely *relative* phase of the probability amplitudes for different states, is crucial in determining direction in quantum evolution. It determines the direction of population transfer between two states of the system. As well as the the direction of emission from collective of excited atoms.

5.4.1 Quantum dynamics in full colour: a new convention for figure making

> *I consider that I understand an equation when I can predict the properties of its solutions, without actually solving it.*
>
> P A M Dirac (1902–84)

To get a clearer understanding of this, it is useful to introduce visualization of multi-level system dynamics that completely captures state amplitudes and their phases. To start, we consider a two-level system consisting of ground $|g\rangle$ and excited $|e\rangle$ state. In rotating wave approximation we can describe coherent driving with resonant Rabi frequency Ω, detuned by Δ from the transition resonance, with the following Hamiltonian

$$\mathcal{H} = \begin{pmatrix} 0 & \Omega/2 \\ \Omega^*/2 & \Delta \end{pmatrix}. \tag{5.12}$$

Notation for the figure is as follows:
- System **states** are represented as grey circles instead of the usual lines.
- State **probability amplitude module** is represented with a coloured circle inside representing population: probability amplitude is proportional to the radius of the circle, and when equal to 1 it matches the radius of the initial grey circle. Surface area of the circles can be used to estimate relative probabilities for different states.
- State **probability amplitude phase** is represented with a colour, which maps phase based on the colour wheel introduced. As a redundant mechanism to readout amplitude and phase, that does not depend on colour recognition, a small grey circle is represented where normally a tip of phasor would be.

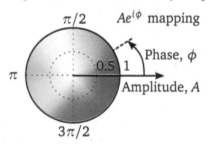

- Calculated **driving strength and phase** is represented as two circle edges: the external one representing driving up in energy (corresponding to $-i\Omega$ term) and the internal one representing driving down (corresponding to $(-i\Omega^*$ term). Colour and encircled semi-arch are consistent with the phase that probability amplitude acquires going up and down (respectively).
- Relative **state energies**, remaining after going to rotating frames, will introduce additional rotation of the state probability amplitudes, with corresponding angular frequency (in the case of the two-level system above, Δ).

With these rules, for any number of quantum states coupled with arbitrary Hamiltonian we can at any point of time *calculate* a representation that maps exactly this information to the figure[11]. And following simple rules we can also

---

[11] **HANDS-ON** Access quantum state visualizations online without installation on Science Web Services via https://sws.labbricks.com/qstate Alternatively, install locally Python package from `pip install interactive-publishing` and explore your own quantum states and system dynamics.

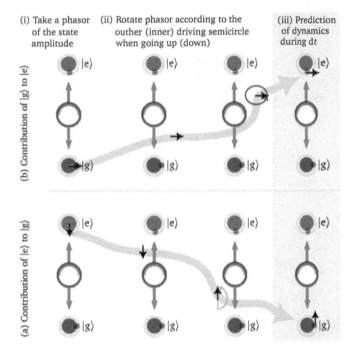

(i) Take a phasor of the state amplitude   (ii) Rotate phasor according to the outher (inner) driving semicircle when going up (down)   (iii) Prediction of dynamics during dt

(b) Contribution of $|g\rangle$ to $|e\rangle$

(a) Contribution of $|e\rangle$ to $|g\rangle$

Figure 5.16. Reading quantum dynamics from new representations of states and Hamiltonian. For a given driving between two state systems and given initial states we can predict the change of dynamics of both ground (a(i)–(iii)) and excited state (b(i)–(iii)) in three quick steps. The chosen example shows atoms state $(|g\rangle + |e\rangle)/\sqrt{2}$, when the laser field suddenly jumps in phase by $\pi/2$ relative to the initial driving phase. Note that this laser phase change cannot be taken under global phase, since it is phase change *relative* to initial driving. Based on two simple steps described in the figure, we can predict that populations of the two states won't be changing, since the new phasor will be adding contribution orthogonality to the existing state. However, a new phasor contribution at 90° to the existing states will make both states accumulate extra phase as driving continues.

predict what happens next, simply using an image to help us with visual integration of a differential system of the equations, as shown in figure 5.16. We will now explore several phenomena, which for two-level systems are not visible in Bloch sphere representation, or, for systems with more than two levels, cannot even be mapped on that simplified representation.

Minus sign from full Rabi flop. With these few simple rules about drawing energy diagrams of the system, we can provide a much more accurate representation of system dynamics than is usually done. Starting with a two-level system (figure 5.17) for resonant driving $\Delta = 0$ we see how going to the excited state and back does not return the two-level system exactly to the initial state. Instead, now there is an extra minus sign acquired as a consequence of twist by driving[12].

---

[12] Note that this is not visible in the standard Bloch sphere state representation of the two-level system, because such representation is not complete, missing phase information relative to some fixed initial phase.

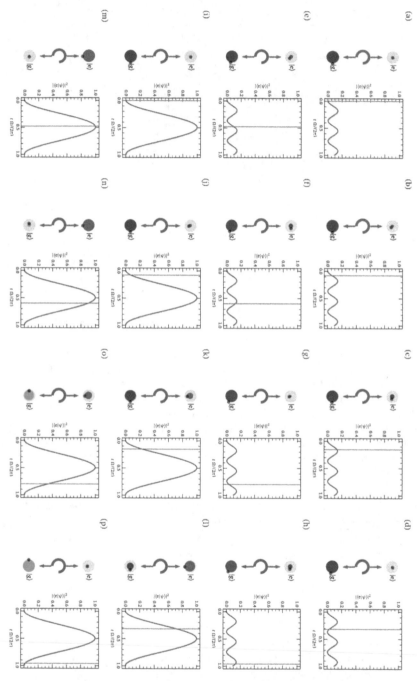

Figure 5.17. Driving two-level system with different drive detunings: each panel shows one moment, indicated in the population plot with a vertical grey line, and a corresponding level diagram with state and driving. Driving detuning: (a-h) $\Delta = -15.7\Omega$, (i-p) $\Delta = 0$. An interactive figure is available at http://doi.org/10.1088/978-0-7503-2628-5.

Faster off-resonant oscillations and extra phase shifts. For off-resonant driving, the excited state twists into the opposite phase every π/Δ, which means that population transfer reverses its direction before full inversion, and that oscillations up and down are happening at a faster rate $\sqrt{\Omega^2 + \Delta^2}$. In addition to speeding up of period of oscillations away from resonance, we also see (figure 5.17) how this off-resonant driving leads to additional phase rotation of ground state, adding effectively energy offset of the ground state whose direction depends on the sign of the detuning (this is so-called **AC Stark shift**). We used such an energy shift brought by off-resonant driving earlier, in section 5.2.4, for non-demolition measurement.

Phase matching. Note that the above described situation holds for say two-level atoms but a mathematically analogous situation occurs when two light field modes are coupled through evanescent wave coupling, like in two evanescent wave coupled waveguides (figure 5.18), or through nonlinearity of the medium with which they interact. We can take as a time axis propagation distance in the medium, starting with initially populated one mode. Due to the mentioned coupling, the probability amplitude describing the population of the other mode will start growing, and it will oscillate between the two modes. However, the analogous effect of detuning Δ is here played by the mismatch $k_1 - k_2$ between wave vectors for propagation of the field in the two modes. So after some distance, the two amplitudes will start to be out of phase. If wave vectors are matched $k_1 - k_2 \to 0$ or they are changing along the propagation distance in the crystal in such a manner that on average $\langle k_1 - k_2 \rangle_L \to 0$ $[L \ll \pi/(k_1 - k_2)]$ as in periodically-poled nonlinear crystals, after some propagation distance the initial mode will be empty and the other mode will be populated. If we continue propagating with two modes coupled, there will be oscillations between the two modes, as in resonant Rabi oscillations. If there is phase mismatch, there will never be complete transfer of the initial field in the second mode, with oscillations happening at propagation distance $\sim \pi/(k_1 - k_2)$. This issue of **phase matching** comes time and again in nonlinear optics. Alternative ways of addressing phase matching are using propagation of two modes in birefringent crystals, whose propagation direction can be at an angle to the crystal axis, and thus allow for some tunability via tilt angle, and by using temperature to tune crystal optical propagation properties.

Driving field phase. Before considering more complex situations, let's explore the effect of the phase of the driving field on the state dynamics. In figure 5.19 we can abruptly change phase of the driving field[13]. A rather dramatic example of seeming *dynamics freezing in spite of continuous driving* is obtained if we change driving phase midway during through the Rabi-flip by $\pi/2$ where the driving laser will be adding contributions to the probability amplitudes of ground and excited state that are at 90° to the existing probability amplitudes. In other words, laser driving will not change populations that are obtained by that moment in two states, but it will continue just *rotating* phase of the two states[14]. On the other hand, if the change of

---

[13] For optical driving fields, like laser, this is typically done using rapid phase switch in radio frequency driving of electro-optic (EOM) or acousto-optic modulator (AOM) through which driving field propagates. Modulated portion of the light will inherit that RF-induced phase shift.

[14] This is also not visible in the standard two-level Bloch sphere representation.

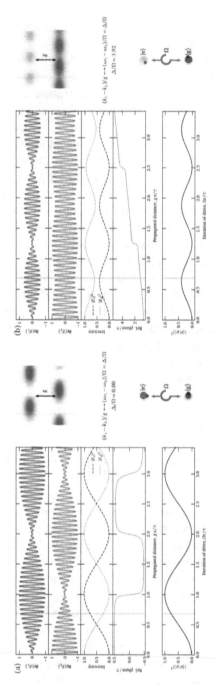

Figure 5.18. Phase matching between light modes versus tuning to resonance driving field of a two-level system: amplitude, intensity and relative phase (top left) of light fields in two waveguide light modes that are coupled via evanescent wave (top right). The dynamics is analogous to that of the population (bottom left) of a driven two-level system (bottom right) (a) phase matched $k_1 = k_2$ case corresponding to resonant $\Delta/\Omega = 0$ driving and (b) phase mismatched coupling of two modes corresponding to off-resonant driving $(k_1 - k_2)\Omega/g = \Delta = 3.92$.

In the phase mismatched case (off-resonant driving) population in one state accumulates extra phase relative to the other as visible on the relative phase drift. Once this extra accumulated phase is more than $\pi/2$, direction of the light (population) transfer direction is reversed, leading to faster overall oscillations of the intensity (population) and lack of complete transfer between two light modes (system states). Some of the common approaches for phase matching two light modes in solid material include using temperature tuning adjusting wave vector, tilting if medium is not isotropic, or stacking of segments with slightly larger and slightly smaller wave vector, so that phase slipping is not allowed to go over $\pi/2$ as in periodically-poled crystals. An interactive figure is available at http://doi.org/10.1088/978-0-7503-2628-5.

phase of the driving field is π, we will *invert direction* of dynamics. Check phases of vectors added by driving to the ground and excited state, contributed by the excited and ground state amplitudes, respectively, picked and rotated by the corresponding driving. You will see that direction of the added amplitudes relative to the existing state amplitudes[15] is also then in π offset, reducing the thus-far accumulated amplitudes. This is often used in coherent manipulation of nuclear magnetic spins also, that are driven by radio frequency fields directly, where the sequence is known as **rotary echo** nutation nuclear magnetic resonance (NMR).

Note that in all the above discussion, what established a well defined phase between the two states of the system were interactions. Now the global phase, such as the initial ground state phase or laser phase, did not matter. It can be fixed at some point and factored out. What matters are changes in *relative* phases between states, as well as between initial and current driving fields, from that point onward.

5.4.2 Measurement induced feedback with no decay leading to fixed relative phase: electromagnetically induced transparency transient example

One way well defined phase can arise is through the measurement induced back action. To see this, we will now unite previously discussed measurement induced feedback in section 5.3 with the newly introduced dynamics visualization from section 5.4.1. We will explore a three-level system with energy levels for Λ structure (figure 5.20). Two states, $|a\rangle$ and $|b\rangle$ are stable while the third excited state $|m\rangle$ can spontaneously decay to either of the two lower states. There are two driving fields Ω_{am} and Ω_{bm} that are resonantly driving transitions a $\leftrightarrow$ b and b $\leftrightarrow$ m transitions, respectively.

The steady state of this system is known to exhibit a special situation where in spite of driving both transitions to $|m\rangle$, that state remains unpopulated, and hence it does not fluoresce. It forms a so-called **dark state** in the steady state, and since without population transfer driving fields are not absorbed, that dark state is also the origin of what in the transmission spectrum of the driving light is seen as transparency window in spite of driving being resonant with the atomic transition. This is so-called **electromagnetically induced transparency** (EIT). The goal of this section is to understand how this state arises in the initial system which is in ground state $|a\rangle$.

---

[15] We are doing this in the spirit of standard phasor additions, just now using visual elements to do quick calculations and understand what is happening.

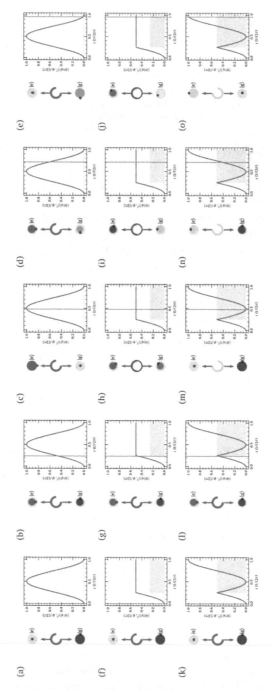

Figure 5.19. Changing phase of field driving two-level system: (a)–(e) resonant driving with fixed phase. After the 2π driving pulse ground state accumulates π phase shift. (f)–(j) After $\pi/2$ pulse driving field phase changes by $\pi/2$. Populations of the two states will not change, but driving will be rotating their respective phases. (k)–(o) After $\pi/2$ pulse, driving changes phase by π, reversing the direction of the population transfer. An interactive figure is available at http://doi.org/10.1088/978-0-7503-2628-5.

(a)

$$i\hbar \frac{\partial |\psi\rangle}{\partial t} = \begin{pmatrix} 0 & \Omega_{am} & 0 \\ \Omega_{am} & 0 & \Omega_{mb} \\ 0 & \Omega_{mb} & 0 \end{pmatrix} |\psi\rangle$$

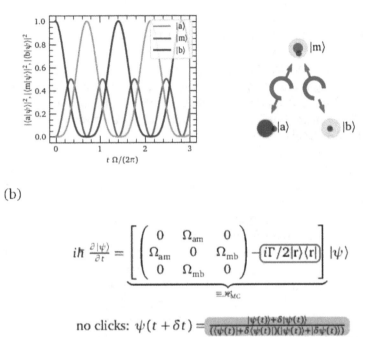

(b)

$$i\hbar \frac{\partial |\psi\rangle}{\partial t} = \underbrace{\left[\begin{pmatrix} 0 & \Omega_{am} & 0 \\ \Omega_{am} & 0 & \Omega_{mb} \\ 0 & \Omega_{mb} & 0 \end{pmatrix} - \boxed{i\Gamma/2 |r\rangle\langle r|} \right]}_{\equiv \mathscr{H}_{MC}} |\psi\rangle$$

no clicks: $\psi(t+\delta t) = \dfrac{|\psi(t)\rangle + \delta|\psi(t)\rangle}{\sqrt{(\langle\psi(t)| + \delta\langle\psi(t)|)(|\psi(t)\rangle + |\delta\psi(t)\rangle)}}$

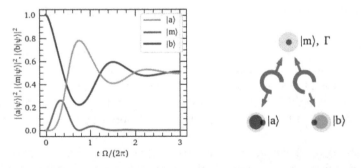

Figure 5.20. Reaching steady state EIT: (a) with no decay from intermediate state $|m\rangle$ steady state is never reached. (b) With existing, *but not-observed*, decay from intermediate state, due to state renormalization under no-click dynamics, steady state is reached, where excitation amplitudes of $|m\rangle$ from $|a\rangle$ and b$\rangle$ destructively interfere since they are exactly out of phase. An interactive figure is available at http://doi.org/10.1088/978-0-7503-2628-5.

First, let us understand how it is possible that two fields are coupling ground states to the excited state and yet the excited state remains non populated. To see that, observe the atomic state in the steady state (at long times) when decay of excited state is present. The probability amplitudes of states $|a\rangle$ and $|b\rangle$ will have opposite phases, and such modulus that, multiplied by the strengths of two driving fields, their two contributions to the excited state at each time step are exactly the same in magnitude but pointing in opposite directions. Hence, there are two paths for populating the excited state, but due to *destructive interference* between them both, there will be no change of population of the excited state. Now we see clearly phases of the driving fields also in this picture, it becomes evident that for experimental demonstrations of such an EIT phenomena relative phases of the two driving fields have to be fixed[16].

Now we understand what the steady state would look like, the question is how it is reached starting from initial conditions. We will explore now transients in two situations: (i) if there is no decay from state $|m\rangle$; and (ii) if there is finite decay rate $\Gamma = \Omega_{am}$ from state $|m\rangle$. Without decays, we see that we just have Rabi oscillations between the three levels. They will be interrupted of course from time to time by the decay from the excited state, but then we just reinitialize the system in either state $|a\rangle$ or state $|b\rangle$, continuing in between with Rabi oscillations. The system never reaches steady state in this situation. On the other hand, if we turn on decay from the excited state, again there are situations where we start with Rabi driving, and then decay from excited state resets us to one of the two lower-lying states. But the interesting dynamics happens when *no decays happened* in spite of the excited state having finite lifetime, thus being effectively observed by the environment. As a consequence of the state renormalization, to adhere to the rule that a state has to be consistent with the total retrieved information about the system, the system state is steered towards the dark state. It is precisely the measurement induced back action, in the absence of measurement events (clicks), that enables reaching this steady state, with two lower states having precise ratio of amplitudes and relative phases.

5.4.3 Well defined relative phase, with no global phase: example of two inverted two-level systems

In section 5.4.2 we've seen how well defined phase between system states can be established as a consequence of continuous measurement back action when no quantum jumps are occurring. Now we will consider how a well defined relative phase can be brought by quantum jumps. And it will be special because the global phase not only won't be important, as argued previously, but will be actually not defined.

Consider a two-level atom with ground $|g\rangle$ and excited $|e\rangle$ state. If we apply so-called π pulse, we will completely invert its population into the excited state $|\psi\rangle = |e\rangle$. The field from the single inverted atom doesn't have a well defined phase.

---

[16] This is experimentally achieved typically either by using ultra stable locked lasers, or by deriving both driving fields from a single laser that acts as a master oscillator.

It's a single photon and moreover the electric field operator $\langle\psi|E|\psi\rangle = 0$ has zero expectation value. But it will decay, and photons will be detected $\langle\psi|E^\dagger E|\psi\rangle \propto \exp(-\Gamma t)$, and will have characteristic exponential decay distribution, that can be obtained if we repeat the experiment many times and bin arrival times are relative to the initial excitation by the π pulse.

Now let's assume that we have two two-level atoms, labeled with index 1 and 2. The second atom has, maybe by being of a different type, slightly different energy of the excited state, offset by $\hbar\Delta$ relative to the first atom. We initially excite both atoms by a broadband pulse that prepares them in a doubly excited state $|e_1e_2\rangle$. The fluorescence from both atoms is combined and mixed on the beam splitter, before being detected on two single-photon detectors, as shown schematically on figure 5.21. Due to the overlap of information caused by the beam splitter, we cannot differentiate between a decay in one or the other atom. Hence, after the first click is detected, the system is in superposition state $|e_1g_2\rangle + |g_1e_2\rangle$. The second photon, corresponding to the decay of the remaining excited state, will be emitted at some time δt after the first photon. At the point of just before the emission, the wave function of the excited state $|e_1g_2\rangle + \exp(i\Delta\ \delta t)|g_1e_2\rangle$ will acquire relative phase $\Delta\ \delta t$ between the two possible paths due to mismatch in energies of the two excited states.

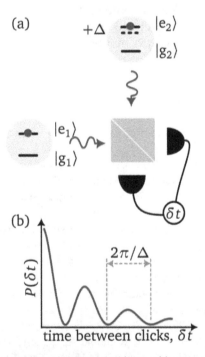

Figure 5.21. (a) Two initially inverted two-level systems decay with same rate Γ, and their fluorescence is combined on the beam splitter before being detected on the photon counter which measures delay δt between the two photons arriving. Two systems are identical apart except for the one excited state $|e_2\rangle$ being $\hbar\Delta$ higher in energy compared to $|e_1\rangle$ state. (b) Probability for second photon detection δt after the first photon detection will show oscillations (beats) in exponentially decaying envelope $\propto\exp(-\Gamma t)$.

Probability $P(\delta t)$ for detection of the second photon at time δt after the first photon will be

$$P(\delta t) = \left| \left\langle g_1 g_2 \right| \left(|g_1\rangle\langle e_1| + |g_2\rangle\langle e_2| \right) \left(|e_1 g_2\rangle + \exp(i\Delta\ \delta t)|g_1 e_2\rangle \right) \right|^2 \exp(-\Gamma t)$$

$$\propto \cos^2\left(\frac{\Delta\delta t}{2}\right)\exp(-\Gamma t).$$

Thus, if we repeat the experiment many times, and measure $g^{(2)}(\delta t) \propto P(\delta t)$ of arrival times of the second photon relative to the first, we will obtain oscillations. These oscillations are **quantum beats** that originate from the *relative* phase between two possible photon decay events that add up coherently to the final photon detection probability amplitude, and interfere. This is happening in spite of the fact that the phase of the field produced in a decay of any individual excited two-level system is undefined. What is important is the relative phase between the paths that contribute to the final decay amplitude, and that *relative* phase is well defined.

To conclude, there are many situations where the relative phase is all that exists. For example, if we are to understand how a single atom can absorb a photon, what happens is really that the single photon wave packet while being scattering on the atom is interfering with its earlier self: the atom provides temporary storage for the front of the photon wave packet, and allows it to contribute to total field later, when the front of the photon wave packet will be coming. A photon interferes with its own front wave packet that was retained by the atom[17].

References

[1] Zurek W H 2002 Decoherence and the transition from quantum to classical—revisited *Los Alamos Sci.* **27** 2
[2] Mollow B R 1969 Power spectrum of light scattered by two-level systems *Phys. Rev.* **188** 1969
[3] Carmichael H J and Walls D F 1975 A comment on the quantum treatment of spontaneous emission from a strongly driven two-level atom *J. Phys.* B **8** L77
[4] Carmichael H J and Walls D F 1976 A quantum-mechanical master equation treatment of the dynamical Stark effect *J. Phys.* B **9** 1199
[5] Sargent M, Scully M O and Lamb W E 1974 *Laser Physics* (New York: Addison-Wesley)
[6] Duan L-M, Lukin M D, Cirac J I and Zoller P 2001 Long-distance quantum communication with atomic ensembles and linear optics *Nature* **414** 413
[7] Kimble H J 2008 The quantum internet *Nature* **453** 1023
[8] Dicke R H 1954 Coherence in spontaneous radiation process *Phys. Rev.* **93** 99
[9] Wiesner S 1983 Conjugate coding *ACM SIGACT News* **15** 78
[10] Šibalić N and Adams C S 2018 *Rydberg Physics* (Bristol: IOP Publishing)
[11] Minev Z, Mundhada S, Shankar S, Reinhold P, Gutíerrez-Jáuregui R, Schoelkopf R J, Mirrahimi M, Carmichael H J and Devoret M H 2019 To catch and reverse a quantum jump mid-flight *Nature* **570** 200

---

[17] An interesting experiment that splits a single photon in an interferometer and whose phase is then shifted in one branch of the interferometer is reported in [13].

[12] Feynman R P 1985 QED: The Strange Theory of Light and Matter (Princeton, NJ: Princeton University Press)

[13] Specth H P, Bochmann J, Mücke M, Weber B, Figueroa E, Moehring D L and Rempe G 2009 Phase shaping of single-photon wave packets *Nat. Photonics* **3** 469

Chapter 6

Fighting environmental noise and imperfections

Real world experiments are never perfect: uniform fields are not quite uniform, the background is not quite static, homogeneous samples are not quite homogeneous. To coax out the underlying quantum dynamics such that they manifest clearly in experiments, a number of approaches have been developed. Three of them—Ramsey interferometry, dynamical decoupling, and spin echo, upon which many varieties have been built—are discussed in this chapter.

The methods discussed here are not strictly quantum: i.e. they don't always depend on entanglement and measurement. Rather, they are general approaches that can be used when continuous dynamics[1] is happening, while unknown but

 LabBricks literature graph for this chapter can be found at https://labbricks.com/#11/k37hhpyhid

---

[1] In contrast to *discrete* dynamics presented by, for example, rate-equation models, where a system has some probability to instantaneously switch from one state to another.

relatively stable[2] external influences are contributing in some linear manner to the dynamics. Under continuous quantum dynamics—in the absence of quantum jumps—the above conditions often can be satisfied, lending these methods to widespread use.

6.1 Ramsey sequence

The original goal of the Ramsey sequence was more accurate measurement of transition frequency between two discrete energy levels, for applications such as atomic clocks. Some of the first experiments measuring transition frequencies in coherent manner were done in the 1930s and 40s, when then new radio and microwave frequency sources were starting to be used for direct coherent manipulation between energy levels of atoms and molecules[3]. The pioneer in this field was Isidor Isaac Rabi (1898–1988) and the most direct manner for measuring the transition frequency is named after him: a **Rabi scheme**, also called a single-oscillating field scheme. Here driving is applied for extended time and at the end probability for atoms or molecules to be in an excited state is measured. The probed atomic and molecular beam was typically obtained from a thermal cloud following selection of velocity v and insertion into the vacuum tube. These early experiments were not using pulsed driving, rather the atoms were traveling through the region in space of length L where driving was applied, experiencing pulsed driving *in their frame of reference*, during $t = L/v$. Depending on detuning Δ of driving field oscillations, and intensity of the driving given by Rabi frequency Ω, the probability that atoms are in excited state after traveling through the apparatus is

$$P_e \propto \sin^2\left(\frac{L}{2v}\sqrt{\Omega^2 + \Delta^2}\right).$$

For weak driving, a resonance peak is obtained for $\Delta = 0$ and $\frac{L}{2v} = \pi/2$. The measurement will be more precise the sharper is the detected population versus $d\frac{L}{2v} = \pi/2$ drive detuning Δ exhibits a sharper peak. The brute-force solution would be to just increase the length L of the interaction region. However, this is limited by the maximal region in which external bias fields (in the original experiments' magnetic fields) can be kept uniform. For large L it is hard to keep bias fields uniform and peak broadening is observed, limiting precision improvement.

Faced with this problem, Norman Foster Ramsey (1915–2011) found a solution that was inspired by the approach in astronomical observations like Michelson

---

[2] The methods discussed here are effective as long as unknown external contributions are changing on longer, ideally much longer, time scales compared to the typical duration of dynamics for one 'de-noising' cycle by any of the methods presented below.

[3] Before that the experiments looking at internal quantum states were of Stern–Gerlach type, that could separate atomic and molecular beams using field gradients by their internal states, and spectroscopy measurements of hot gases, discharges etc.

stellar interferometer[4]. In a stellar interferometer, one can increase angular resolution of the telescope by about a factor of two if one takes light only from the edges of the optical telescope, effectively by painting the middle of the telescope in black[5]. The optical quality of the surface of the telescope painted in black doesn't matter—in fact, it doesn't even have to be there. Motivated by this observation, Ramsey looked for an analogous method that would not be susceptible to all the inhomogeneities of the applied bias field that occur in long beam apparatus.

The sequence that Ramsey found—now called Ramsey sequence, involved two regions with oscillating first. In the first short region particles would be driven by $\pi/2$ so that they end up in superposition of the ground and excited state. They would then evolve for a longer time L/v of flight through the part of the apparatus where there was no applied field. Finally, at the end of the apparatus, the second short oscillating field region would drive particles by another $\pi/2$ pulse, and the number of excited particles would be measured. Importantly, both oscillating regions have to have the same driving phase, which is usually achieved by deriving the drive from the same oscillator. Probability of excited state now depends on drive detuning Δ relative to the transition resonance frequency as

$$P_e \propto \cos^2\left(\frac{\Delta L}{2v}\right)$$

We see that resonance P_e versus Δ will be sharper for longer L, and importantly the method is insensitive to the inhomogeneities to the bias field in the region of the length L, only the average bias field matters. The Ramsey sequence is somewhat analogous to the Mach–Zehnder optical interferometer, whereby instead of splitting photons into spatial paths, we split the internal state of the particles into superposition of two paths for internal dynamics, and then recombine them later.

The Ramsey sequence and quantum coherence. The Ramsey sequence has developed into a key methodology in the measurement of quantum coherence and hence decoherence. As the essence of quantum optics is interference and quanta—wave and particle—the Ramsey sequence has taken on a central role in both quantum optics and quantum computing. For example, when we perform a measurement on a two-state quantum system, we are not sensitive to phase information and hence learn nothing about coherence. The Ramsey sequence plays the analogous role as interferometry in optics, allowing us to measure phase as well.

The Ramsey sequence and interferometry. To measure phase we need to perform an interference experiment, where we split the wave into at least two parts. In quantum optics, for single photons we can split the wave (using a beam splitter) but not the particle. In the Bloch or Poincaré picture, if we are with the $|0\rangle$, interference

---

[4] For more details, see interview with Norman Ramsey, that is part of the American Institute of Physics, Oral History program http://www.aip.org/history-programs/niels-bohr-library/oral-histories/31413-3. Similar ideas for increasing angular resolution of relatively bright objects are utilized today under the name **aperture masking interferometry**.

[5] This of course comes at the cost of reduced total received signal, so the technique can be used only with relatively bright sources.

corresponds to an x or y rotation to create a superposition of the two components $|0\rangle$ and $|1\rangle$ (like a Hadamard implement using a beam splitter), a z rotation to change their relative phase following by another x or y to undo the superposition.

What happens when we recombine the two paths? Two possible ways to recombine the paths are illustrated in figure 6.1. Figure 6.1(a) is known as a *Michelson interferometer*. Alternatively, we can recombine them on a second beam splitter as in figure 6.1(b). This is known as a *Mach–Zehnder interferometer*. This has the advantage that the two inputs and the two outputs are all separate. There is a strong analogy between single-photon (or classical) interferometry in real space and interferometry in state space (Hilbert space)—known as *Ramsey interferometry*. Real and Hilbert space interferometry are exploited in photonic and matter quantum computing, respectively. As they are closely related we shall discuss both together in this section.

What happens when the two paths recombine on the beam splitter? Depending on the relative phase between the two paths the photon may be either reflected or transmitted. Figure 6.1(c) shows how the photon signal detected in one of the output modes oscillates as a function of the path or phase difference. This sensitivity to the length of the arms in the Michelson interferometer is exploited in [1].

To model the Mach–Zehnder interference, we use the same two-state model developed in sections 4.9 to 4.14, where in this case the two states are the two paths as in dual-rail encoding of photonic qubits. We aim to find an expression for the probability to detect the photon in one or other of the output ports as a function of the relative phase. Now that we have a matrix to describe the beam splitter we can put one after another to form the Mach–Zehnder interferometer shown in

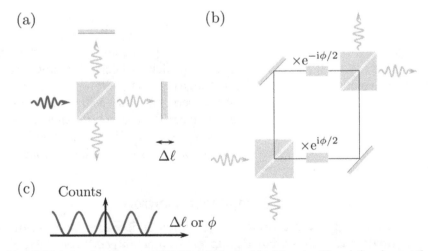

Figure 6.1. (a) Michelson interferometer. (b) Mach–Zehnder interferometer. The relative phase at the second beam splitter, ϕ, may be tuned using an electro-optic medium in one or both arms (c) The photon counts detected in one output port as a function of the path difference, $\Delta\ell$, or phase difference, ϕ, between the two paths. The interference fringes correspond to constructive and destructive interference between the two paths as the relative phase of the two components varies.

figure 6.1(b). Using the Hadamard form of the beam splitter, the transfer matrix for the Mach–Zehnder is

$$
\begin{aligned}
HR_z^\phi H &= \frac{1}{2}\begin{pmatrix} 1 & 1 \\ 1 & -1 \end{pmatrix}\begin{pmatrix} e^{-i\phi/2} & 0 \\ 0 & e^{i\phi/2} \end{pmatrix}\begin{pmatrix} 1 & 1 \\ 1 & -1 \end{pmatrix}, \\
&= \frac{1}{2}\begin{pmatrix} 1 & 1 \\ 1 & -1 \end{pmatrix}\begin{pmatrix} e^{-i\phi/2} & e^{-i\phi/2} \\ e^{i\phi/2} & -e^{i\phi/2} \end{pmatrix}, \\
&= \frac{1}{2}\begin{pmatrix} e^{-i\phi/2} + e^{i\phi/2} & e^{-i\phi/2} - e^{i\phi/2} \\ e^{-i\phi/2} - e^{i\phi/2} & e^{-i\phi/2} + e^{i\phi/2} \end{pmatrix}, \\
&= \begin{pmatrix} \cos\phi/2 & -i\sin\phi/2 \\ -i\sin\phi/2 & \cos\phi/2 \end{pmatrix}.
\end{aligned}
\tag{6.1}
$$

This expression says that for a $|0\rangle$ input, the probability to be in state zero at the output is

$$
P_{|0\rangle} = |a|^2 = \cos^\phi/2,
$$

i.e. the population oscillates between the two output ports as a function of the phase shift ϕ.

Equivalence of interferometry in real space and Hilbert space. In figure 6.2, we show a photon moving through an interferometer. To make the connection with what we have seen before we add both the Bloch sphere representation and the density matrix below. Moving across the rows corresponds to advancing time. Move down the column we change the phase shift, ϕ. We assume that the phase shift is induced by an optical component with a tunable refractive index in each path. Higher index is represented by darker blue. The size and colour of filled circles plotted on top of the interferometer track correspond to operator amplitude and phase, respectively.

Although single-photon interferometry involves both counting and interference, we could watch the interference pattern build up click-by-click. Consequently, there is nothing particularly quantum yet. Even if we use a single-photon source, the result is the same as in the G I Taylor double-slit experiment with feeble light in 1919 (see Discovery of Quantum Optics section, chapter 2). The interference fringes are exactly the same as predicted by classical optics, see e.g. [2]. The missing ingredient in going from classical to quantum is what happens when we add a second quanta. This will be discussed in the next chapter.

6.2 Dynamical decoupling (bang–bang control)

The objective of dynamical decoupling is prevention of decoherence of the quantum state of a single quantum system due to environmental perturbation. In many situations there is some quantum information encoded in the state of a system—for example, a two-level system can encode a so-called qubit of information. However, this information can effectively be erased by the process of decoherence: for example one of the two states can couple to a generally unknown environment more than the

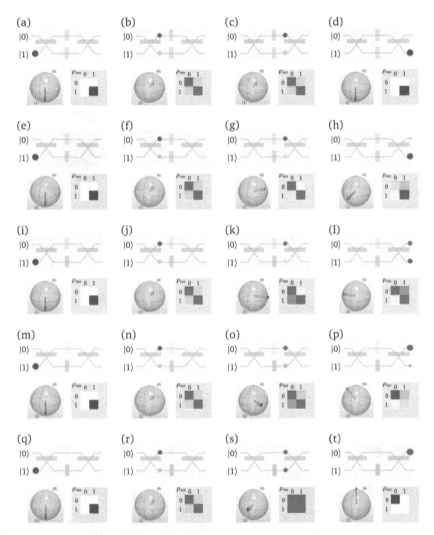

Figure 6.2. Photonic circuit version of the Mach–Zehnder interferometer shown in figure 6.1(b). The beam splitters and glass plates implement $HR_z^\phi H$ sequence expressed in equation (6.1). This type of Mach–Zehnder interferometer is a key building block in photonic quantum information processing. The rows show different times as the photons move through the interferometer. The columns (parameter slider) correspond to increasing the phase, ϕ: (a)–(d) $\phi = 0$, (e)–(h) $\phi = \pi/4$, (i)–(l) $\phi = \pi/2$, (m)–(p) $\phi = 3\pi/4$, and (q)–(t) $\phi = \pi$. As a function of the phase, ϕ, the output oscillates between $|1\rangle$ (top row) and $|0\rangle$ (bottom row). In the bottom row the output is $-i|0\rangle$ (represented by blue in colour wheel) as expected from equation (6.1). Below we show the Bloch vector and density matrix corresponding to the state vector at each position in the circuit. In section 7.10 we shall discuss how it is employed to realize an all-optical two-qubit gate. An interactive figure is available at http://doi.org/10.1088/978-0-7503-2628-5.

other state, and experience some extra phase shift. Even if there is no relaxation (dissipation) of the system, that coupling to the environment would mess up the relative phase between the two excited states, deleting part of the stored information in the qubit.

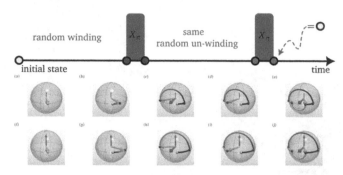

Figure 6.3. Bang–bang dynamical decoupling. Timeline of decoupling sequence and state evolution shown on Bloch spheres for weak (a)–(e) and strong (f)–(j) external perturbation that introduces additional state dephasing over time equivalent to the rotation around the red arrow directed along the z-axis. Sequence from left to right: initial state (a,f), just before first X_π pulse (b,g), just after first X_π pulse (c,h), just before second X_π pulse, and just after second X_π pulse. An interactive figure is available at http://doi.org/10.1088/978-0-7503-2628-5.

If the environmental perturbation is not changing very slowly, staying essentially the same over time needed to apply multiple coherent control drive pulses, one can engineer a driving sequence such that at the end the sequence the environmental contribution to evolution averages to zero. That is the idea behind dynamical decoupling.

In the simple case, switching population between two states by applying π pulses (figure 6.3), ensures first that both probability amplitudes constituting two-level state accumulate the same extra phase due to second order perturbation after the first π pulse, and then removes this extra phase by switching sign of the newly contributed phase after second π pulse.

6.3 Spin-echo

The objective of spin-echo sequences is measuring ensemble response over time in spite of inhomogeneity of the medium that perturbs different parts of the system. In principle, dynamical decoupling sequence from the previous section on each member of the ensemble would maintain state insensitivity to inhomogeneous dephasing over all time. However, there is a simpler sequence available if we just want to rephase the ensemble at some time later T after initial driving.

The idea is to let the system evolve for half of time $T/2$ under an inhomogeneous environment, dephasing each member of the ensemble with a different rate dependent on local perturbation (figure 6.4). Then a π pulse is applied (in the equator of the Bloch sphere but perpendicular to the initial drive), setting each state vector to the transverse back path that it transversed during the first $T/2$. As long as the inhomogeneous environment stays static on timescale T, dephasing rates for each member of ensemble remain fixed, and at time $T/2$ after π pulse, all states return to the same initial state. This is analogous to a marathon where all participants run at their own speed, but half-way through they reverse direction, and start running back with the same speed. As long as individual speeds don't change, everyone arrives

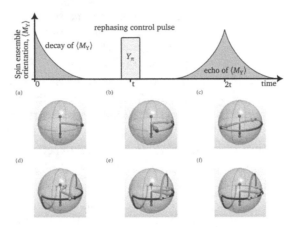

Figure 6.4. Spin echo sequence. Due to different local fields, different spins in the system experience different state energy offsets. During evolution, their state vectors will evolve differently, leading to depolarization of the collective spin ensemble. Bloch spheres show trajectories of states (blue arrows) for two such spin (red and blue arrow tips), each evolving under influence of their own state detunings (gray arrows). If after some time t we apply a Y_π pulse, and wait for t more, all spin states, independent of the exact local energy offset they experience, will reach the initial point in the Bloch sphere. This rephasing of the spin orientations is called **echo**. An interactive figure is available at http://doi.org/10.1088/978-0-7503-2628-5.

back to the starting point independent of their running speed. Return of all members of the ensemble into the initial state is so-called **echo**. Of course sequence does not protect from dissipative decay that might be happening with rate T_1. Size of the echo signal after different waiting times T enables determining dissipative decay time T_1, decoupled from the additional inhomogeneous dephasing time T_2.

6.4 Switching external perturbation: Dicke narrowing

All the above examples used driving to switch position of the state vector relative to the perturbation which remained constant. Thanks to a carefully selected switching sequence, the influence of perturbation on the state vector dynamics averages to zero. One might ask if the inverse happens: that we have constant driving, but that perturbation switches position relative to the state vector quickly, just in the right manner to cancel its influence on average state vector dynamics.

Such examples indeed exist. To see this consider a system of two-level atoms in vapour. The atoms are in thermal equilibrium and they fly with velocities v, whose probability distribution $p(v)$ is given by Maxwell distribution $p(v) \propto \exp(-mv^2/2k_BT)$ for atoms with mass m in thermal equilibrium cloud at temperature T. If we try to drive such an ensemble of atoms with a laser field, each atom would experience a Doppler shift of the laser driving frequency of $\Delta = \mathbf{k} \cdot \mathbf{v}$, where $k = \lambda/2\pi$ is amplitude of driving field wave vector, directed along the the driving laser beam. If we take a spectrum of excited state probability versus driving detuning, this normally leads to Doppler-broadened transition. But there are two interesting cases where measured transition spectra resonance is actually much narrower.

Atom-buffer gas non-depolarizing collisions in dense gas. If the atoms are elastically colliding often with some buffer gas—so that in collision atoms maintain internal state but change direction of velocity—what happens in the Bloch picture is that driving Ω that starts somewhere in the equatorial plane, due to Doppler broadening rotates in one direction with rate $\Delta = \mathbf{k} \cdot \mathbf{v}$ in one direction before collision, and then after collision with rate $\Delta' = \mathbf{k} \cdot (-\mathbf{v})$ in the opposite direction. If these collisions are happening at high rate so that during time t driving field vector rotates only by a small amount $tkv \ll 2\pi$, which is equivalent to condition of flying only δx, $\frac{\delta x}{\lambda} \ll 1$ between the collisions, the driving field vector will not have time to move significantly in equatorial plane before changing direction of the motion. On average, small movements of the driving field Δ due to changes in the detuning average to zero. This means that atoms in sufficiently dense buffer gas, where collisions are happening at high enough rate that they move much less than λ between collisions, evolve under driving as if there is no Doppler shift, and associated motional broadening of the transition line in the spectra. This phenomenon is called **Dicke narrowing**[6].

A related effect to Dicke narrowing happens with nuclear magnetic resonance, where a diffusing atom that travels through some medium experiences continuously changing perturbation due to changes of the local environment. If this happens at the rate faster than the driving, atom dynamics will feel only averaged perturbation from the changing environment. This leads NMR spectral lines to be much narrower in real cases of atoms *moving* through diffusion through the medium, than what would be naively obtained from the ensemble of *static* atoms in the inhomogeneous medium. This effect is called **motional narrowing**[7].

In ultra thin vapour cells. Finally, we draw attention to another effect that causes much narrower lines than expected from a Doppler-broadened driven atomic ensemble. It happens in very narrow vapour cells, where separation between two flat glass panels between which atomic vapour is held is only a few driving wavelengths λ thick. Here also, narrowing and broadening of resonance is observed with period λ experimentally. However, the conditions here are different compared to the original Dicke narrowing explored above, because now there is no buffer gas, but just very narrow spacing between the two glass plates where atoms can fly before colliding with the wall. Importantly, and in contrast to buffer gases chosen for Dicke narrowing demonstrations, these collisions completely *depolarize* atoms, effectively rotating the state vector in the equatorial plane of the Bloch sphere by random angle at each collision. So coherent averaging of single atom continuous dynamics does not cancel Doppler contribution in consecutive collisions. And yet something does cause specific periodic oscillations of the spectral lines with varying distance between plates in the ultra thin cells.

The reason becomes clear if one considers not single atoms, but *pairs of atoms*. To understand the origin of oscillations in the linewidth, consider two atoms leaving

---

[6] Explained for the first time in [3].

[7] Described for the first time by Nicholaas Bloembergen (1920–2017) in his PhD thesis; similar and related narrowing mechanism is **exchange narrowing**, both of which are essentially 'random frequency-modulation' of transition frequency in the presence of the driving field, see [4].

opposite glass plates with velocities equal in magnitude but oppositely directed. These two atoms experience two different Doppler detunings $\pm \mathbf{k} \cdot \mathbf{v}$ of driving field, so their state vectors rotate—as they fly away from respective glass walls—in opposite directions (figure 6.5). Importantly, if we look at the *sum* of two-state vectors, with coherence proportional to polarization and hence emitted electric field, it will add in phase with the coherence of the non-moving atoms for whom there is no extra Doppler detuning. This is the case only as long as the state vectors of the two atoms have not rotated in the equatorial plane of the Bloch sphere by more than $\pi/2$, which happens after they have traveled time t, $kvt = \pi$, traversing distance $\delta x = \lambda/2$. Then their contribution is exactly zero, and from then on it starts to contribute exactly opposite to the coherence of zero-velocity atoms. Until their states rotated by $kvt = \pi$ during time t, which is equivalent to traversed distance $\delta x = \lambda$, when their contribution is again exactly zero and from then on it starts to add in phase to zero-velocity atoms. Due to this constructive and destructive interference, all velocity groups in pairs, for symmetric velocity distribution like the Maxwell–Boltzmann one, contribute to the resonance peak, first producing constructive interference, raising peak of resonance, and then producing constructive interference with zero-velocity atoms, reducing peak resonance. This interference makes a peak for zero-detuning to oscillate in height, producing effective relative osculations in the peak width, typically calculated as full width at half maximum of the resonance peak.

6.5 Non-Markov environments: coupling initially independent field modes

From the point of view of experimental physicists, wanting to explore quantum dynamics for a long time, the major imperfection brought by the environment of quantum systems is that information leaks to it even when we don't want to do measurements, leading to decoherence and return to classical dynamics. A major mechanism for this dissipation in matter–light interactions is spontaneous emission of atoms (or equivalent optical emitters like quantum dots etc). To tackle and control atom–light interactions, it is important to realize that spontaneous emission rate is *not a property of atoms themselves*, but a property of an atom *and* local field modes. Standard decay rate Γ is derived[8] usually under the assumption that an atom radiates in free space, and that every possible field mode is independent, and Markovian. That is, field modes lose immediately any memory of the past processes, and in this case have no memory that the atom started decaying into them in the past, so that atom experiences the same interaction with empty modes throughout its decay. However, what if that is not the case? What if we have such an environment where the fact that an atom was in an excited state not long ago, and started populating probabilities for non-zero photons in decay modes, has been memorized by the environment.

Coupling field modes and adding memory. Adding memory to an atom's environment can be easily done, for example, by placing the mirror in proximity to the

---

[8] This is the standard Weisskopf–Wigner theory of spontaneous emission.

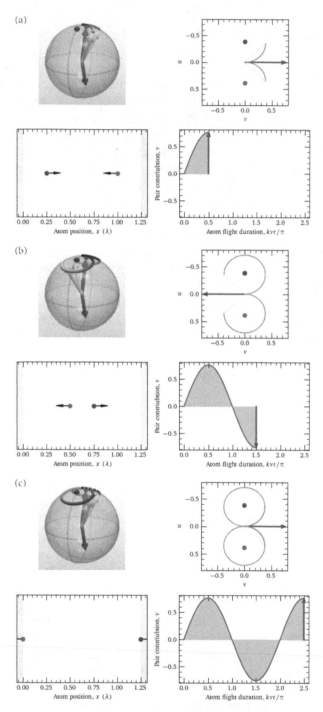

Figure 6.5. Origin of Dicke narrowing of spectra in thin vapour cells. Top right panel: Bloch sphere showing two-state vectors (yellow arrows) corresponding to two atoms traveling with velocities equal in magnitude but opposite directions. The states evolve according to two red and blue driving vectors $(u, v, w) = (\Omega, 0, \pm kv)$, respectively. Dots are intersections of the wave vectors with the north hemisphere. Top right: projection of the

evolution in the (u, v) plane. Starting from $(0, 0)$, two-state vectors (yellow) rotate around tips of the drive vectors (red and blue points). The sum of their contributions to total scattered light is directed along the v axis. Bottom left panel: two atoms with same magnitude of velocity (black) but opposite direction, flying from the cell walls (blue). Bottom right: contribution of the atom pair the scattered light adds constructively regardless of their velocity, for a time t until they transverse $kvt = \pi$. For cells of that thickness $\lambda/2$ we have a peak of the resonant ($\Delta = 0$) spectra. Contribution of all atoms, regardless of their position in the cell, is accounted for by summing the area of the curve above the x-axis. Atom-pair dynamics shown for a cell of total thickness $\frac{5}{4}\lambda$, at times (a) $kvt = 0.5\pi$, (b) $kvt = 1.5\pi$, (c) $kvt = 2.5\pi$ after leaving the cell walls. An interactive figure is available at http://doi.org/10.1088/978-0-7503-2628-5.

atom, at distance d closer than c/Γ (figure 6.6). How can an atom know that there is a mirror close by? Why should its decay change? Of course from the local perspective of the atom, dynamics stays the same, at least initially. Decay starts as usual. There is a finite probability that a photon is emitted in all possible field modes, and this photon 'wave-function' propagates as a wave until it faces the mirror, and then it bounces back partially towards the atom. That means that at time $2d/c$ since the start of the decay part of this photon, probability will *reach back* the atom. At this point the field modes are not Markovian (they do contain memory that the atom started decaying earlier), nor independent (mirror can reflect photon emitted in one field mode into another field mode and return it to the atom). This reflected photon probability interacts now with the atom and can be re-absorbed. The mirror essentially wraps time, allowing a photon to interfere with its earlier self: photon probability originating from possible earlier decay interferes with photon probability originating at the location of the atom from decay at a later time. Depending on the relative phases of these two probabilities, overall decay can be increased or decreased. For understanding of this fundamental effect, it is important to think about the photon probability wave, and not imagine a photon just as a ball. In the words of Milloni and Knight [5], 'The radiated field (or the photon "wave function") contains all this information: it does not represent a buckshot or fuzzy ball photon.'

Direct calculation of feedback fields. Adding a mirror close to the atom is not the only way to couple initially independent field modes, and feed back the memory of the system dynamics back to the system local environment. That can be achieved also by placing at close distance any other surface or resonant emitter. While the above picture of dynamics from the local perspective of a single atom—that now experiences a non-Markovian environment—is useful to understand fundamental processes described in local dynamics terms, for calculating dynamics in practice there are several different approaches. In the case that we have a number of closely positioned resonant emitters, or a few mirrors that we can equivalently model using image charges, we can directly use field propagators and calculate system dynamics. In that way we can model dense dipoles or **Casimir–Polder** and **van der Waals** atom–surface interactions.

New independent local field modes. In the case that enclosing surfaces are complicated, adding many different feedback paths, and coupling together many initially independent field modes as happens in small optical cavities, it is often useful to diagonalize these initially independent field modes to find a set of new

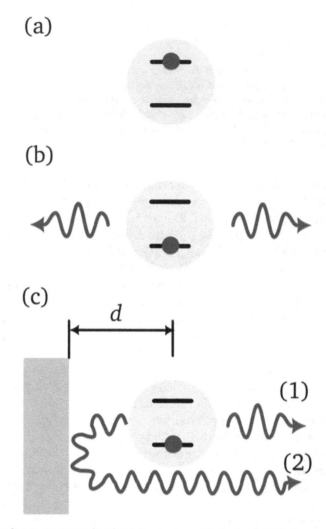

Figure 6.6. Mirror close to an atom makes the environment non-Markovian. Initially inverted atom (a) decays in free space with some rate Γ (b). When placed closed to mirror (c) at distance $d < c/\Gamma$, probability that photon [c (2)] is emitted to the left at some *earlier* moment $t - \delta t$, $\delta t = 2d/c$, and reflected off the mirror, interferes with probability that photon [c (1)] is emitted at moment t. For $d \to 0$, since reflection off the mirror introduces π phase shift, the two possible paths (1) and (2) in dynamics for photon propagating away from the atom are exactly out of phase (relative) and cancel. Instead photon (2) will be re-absorbed bringing the atom to excited state (a), resulting in vanishing spontaneous emission rate $\Gamma \to 0$.

independent fields, that is, to find new local field mode density. These new field modes will in general have different mode volumes, frequencies of their total number compared to initial free-field modes. Reducing the mode volume V increases coupling between dipole and field $g \propto 1/\sqrt{V}$. Combination of stronger coupling, and modified density of modes, can increase or reduce atom–field coupling in cavities, and is usually measured relative to free space coupling through the **Purcell factor** [6]. This is how a regime of strong atom–light coupling can be achieved, which we will discuss in chapter 9. Cavities can be used also to completely remove resonant decay modes and **suppress spontaneous emission** [7]. Note that while the above mechanism works not only for optical fields, but also for general waves, for example, it is also used in mechanical oscillators, to reduce dissipation of them by suppressing unwanted phononic modes.

Finally, while we focused above on well defined individual emitters, modifying local field modes can also influence radiation of more chaotic, collective sources, like that of a black body, as was recently shown by shaping black body radiation with metasurfaces that modify local field modes [8].

References

[1] Castelvecchi D 2017 Gravitational wave detection wins physics Nobel *Nature* **550** 19

[2] Adams C S and Hughes I G 2019 *Optics f2f: From Fourier to Fresnel* 1st edn (oxford University Press) https://doi.org/10.1093/oso/9780198786788.001.0001

[3] Dicke R H 1953 The effect of collisions upon the Doppler width of spectral lines *Phys. Rev.* **89** 472

[4] Anderson P W 1954 A mathematical model for the narrowing of spectral lines by exchange or motion *J. Phys. Soc. Japan* **9** 316

[5] Milloni P W and Knight P L 1973 Spontaneous emission between mirrors *Opt. Commun.* **9** 119

[6] Purcell E M 1946 Spontaneous emission probabilities at radio frequencies *Phys. Rev.* **69** 681

[7] Yablonovitch E 1987 Inhibited spontaneous emission in solid-state physics and electronics *PRL* **58** 2059

[8] Overvig A C, Mann S A and Alù A 2021 Thermal metasurfaces: complete emission control by combining local and nonlocal light-matter interactions *Phys. Rev.* X **11** 021050

Part II

Two or more quanta

IOP Publishing

An Interactive Guide to Quantum Optics

Nikola Šibalić and C Stuart Adams

Chapter 7

Two photons

7.1 Introduction

In this chapter, we aim to show how the addition of a second quanta 'changes everything'. The reason is that with two photons we can have interactions and **entanglement**. Entanglement is particularly important as there is no classical analogue.

There are two ways to produce two photons. First, we can take a single-photon source like in the previous chapter and execute an emission event twice. Subsequently, we can time delay the first photon to make them simultaneous. Alternatively, but more challenging is that we could make two identical single-photon sources. The second method is to make a **two-photon emitter**. Again there are many ways to do this. We can use atoms that exhibit two-photon decays, use non-linear processes such as four-wave mixing or split a photon in two—parametric down conversion—using non-linear optics. In this case, if we want to subsequently interfere the two photons they need to be identical, but to demonstrate entanglement the photons do not need to be identical.

In this chapter we shall discuss examples using identical photons and non-identical photons. First we discuss what can happen when two identical photons interfere. We shall find that **quantum entanglement** emerges naturally in the

LabBricks literature graph for this chapter can be found at https://labbricks.com/#12/6wii9toLFx

two-photon interference process. This is known as the **Hong–Ou–Mandel (HOM)** effect after their seminal experiment in 1988, see timeline. We discuss how two-photon interference can result in an example of the simplest form of two-qubit entangled states, known as a **Bell state**. Next, we consider how Bell states can also be produced using two-photon emitters.

We discuss how the non-separability of entangled renders any representation based on individual quantum systems like the Bloch or Poincaré sphere picture—that was the focus of Part I—useless. Consequently, we need a different visualization tool such as the density matrix. Finally, we consider how two-photon interference can be exploited to realize a **two-qubit quantum gate**.

As in part I, our discussion will follow closely the ideas underpinning quantum information theory. Namely, we can talk about each photon as a qubit which could be either in a superposition of different polarization states (a polarization encoded qubit) or in a superposition of occupying either of two paths or modes (dual-rail encoding). For two photons, we have two qubits and hence the complete range of possible two-qubit states, including entangled states, are accessible. Consequently, we shall devote some time in this chapter to discussing important two-qubit states such as Bell states, and their more bizarre properties. Buiding on these idea in the next chapter we shall discuss a 'classic' experiment demonstrating entanglement, counting and correlation known as the **quantum eraser**.

7.2 Two-photon interference

The most dramatic change between one and two photons is apparent when we consider what happens when two identical photons are incident on a beam splitter. Due to constructive and destructive interference, we find that both photons emerge in the same channel.

Consider the beam splitter discussed in section 4.9—sketched again in figure 7.1(a). If one photon is incident in each path, there are four possible outcomes: both photons transmitted, both reflected, the first transmitted and the second reflected, and the first reflected and the second transmitted. These four outcomes are shown in figure 7.1(b).

Due to the sign change on reflection from a medium with higher refractive index, if the two photons are indistinguishable then cases where both photons are transmitted or both reflected cancel, leaving only the outcomes where both photons are in the same path. This is known as the **Hong–Ou–Mandel (HOM) effect** (see timeline figure in chapter 2 for reference) (figure 7.2).

7.3 Hong–Ou–Mandel effect: Fock basis

In this section, we develop a mathematical description of two-photon interference. We shall focus on the HOM effect for the case of a wave guide beam splitter of the form shown in figure 7.1. If we have two (or more) photons in a particular channel or mode, it is often more convenient to use a *Fock basis*. For two channels, the Fock basis is written as $|n_\mu, n_\nu\rangle_{\text{region}}$, where n_μ and n_ν are the number of photons in channels μ and ν, respectively. The subscript refers to a particular region of space.

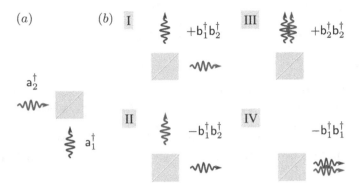

Figure 7.1. (a) Two photons incident on a beam splitter. (b) There are four possible outcomes: I Both photons transmitted. II Both photons reflected. III Photon 1 transmitted and photon 2 reflected. IV Photon 1 reflected and photon 2 transmitted. As there is a sign change on reflection going from lower to higher refractive index, if the photons are indistinguishable then cases I and II cancel, leaving only III and IV. Hence the outcome is a superposition of both photons in one or the other path. This is known as the Hong–Ou–Mandel effect.

Figure 7.2. Hong–Ou–Mandel student lab experiment. To see details of experiment, including alignement of non-linear crystal and tuning of optical path difference for two photons, scan QR code or open https://www.researchx3d.com/?o=9&s=13 directly. The annotated model is based on the setup in the Laboratoire d'Enseignement Expérimental (LEnsE) in the Institut d'Optique Graduate School, Universite Paris-Saclay. [Annotated story opens in web browser online; optionally, R3D marker can also be used to see experiment in Augmented Reality].

For the example shown in figure 7.3, there are two distinct regions, before and after the beam splitter. We label the region before the beam splitter as the input ports, and the region after the beam splitter as the output ports. We define the input and output operators as $i^\dagger$ and $o^\dagger$, respectively. The input and output operators only apply in the input and output regions, i.e. the operators $i^\dagger$ only act on $|n_\mu, n_\nu\rangle_{\text{in}}$ and $o^\dagger$ only act on $|n_\mu, n_\nu\rangle_{\text{out}}$. In figure 7.3 we have also introduced a colour coding for each channel.

Figure 7.3. Two photons incident on a beam splitter. The input and output operators, $i^\dagger$ and $o^\dagger$, only apply in the input and output regions, respectively. Channels 1 and 2 are colour coded red and purple, respectively.

In the simplest example of two-photon interference, the case shown in figure 7.1, two photons incident in two separate channels interfere on a beam splitter. In the Fock basis, we write this as

$$|\psi\rangle_{\text{in}} = |11\rangle_{\text{in}}.$$

We can also write this in terms of creation operators acting on the vacuum state,

$$|\psi\rangle_{\text{in}} = i_1^\dagger i_2^\dagger |00\rangle_{\text{in}} = |11\rangle_{\text{in}}.$$

To find the output state, we make use of the general result that for any linear optics network, we can write the output operators as a linear superposition of the input operators, i.e.

$$o^\dagger = \mathcal{M} i^\dagger,$$

where $\mathcal{M}$ is known as the operator transform matrix. The matrix elements are derived from knowledge of the reflectivity and phase shift of each beam splitter in the network. As we saw for one photon, for a wave guide beam splitter, like in figure 7.3, with intensity reflectivity $\mathcal{R}$, the transfer matrix is,

$$\mathcal{M} = \begin{pmatrix} \sqrt{\mathcal{R}} & \sqrt{1-\mathcal{R}} \\ \sqrt{1-\mathcal{R}} & -\sqrt{\mathcal{R}} \end{pmatrix}.$$

For the special case of a 50:50 beam splitter, $\mathcal{R} = 0.5$, $\mathcal{M}$ is the Hadamard operator. To find the output state, we need to express the input operators in terms of the output operators, i.e.

$$i^\dagger = \mathcal{M}^{-1} o^\dagger,$$

and then apply the appropriate superposition of output operators to the vacuum output state, i.e. for the two-photon input state, with one photon in channel m and one in channel, n, $|\psi\rangle_{\text{in}} = i_m^\dagger i_n^\dagger |00\rangle_{\text{in}}$, the output is

$$|\psi\rangle_{\text{out}} = \left(\sum_\mu \mathcal{M}_{m\mu}^{-1} o_\mu^\dagger \right) \left(\sum_\nu \mathcal{M}_{n\nu}^{-1} o_\nu^\dagger \right) |00\rangle_{\text{out}}.$$

For the wave guide beam splitter in figure 7.3, the transfer matrix is symmetric, $\mathcal{M} = \mathcal{M}^{-1}$, and $i^\dagger = \mathcal{M}^{-1} o^\dagger$ in matrix form is

$$\begin{pmatrix} i_1^\dagger \\ i_2^\dagger \end{pmatrix} = \begin{pmatrix} \sqrt{\mathcal{R}} & \sqrt{1-\mathcal{R}} \\ \sqrt{1-\mathcal{R}} & -\sqrt{\mathcal{R}} \end{pmatrix} \begin{pmatrix} o_1^\dagger \\ o_2^\dagger \end{pmatrix}.$$

The output state is,

$$
\begin{aligned}
|\psi\rangle_{\text{out}} &= \left(\sum_{\mu=1,2} \mathcal{M}_{1\mu}^{-1} o_\mu^\dagger\right)\left(\sum_{\nu=1,2} \mathcal{M}_{2\nu}^{-1} o_\nu^\dagger\right)|00\rangle_{\text{out}}, \\
&= \left(\sqrt{\mathcal{R}}\, o_1^\dagger + \sqrt{1-\mathcal{R}}\, o_2^\dagger\right)\left(\sqrt{1-\mathcal{R}}\, o_1^\dagger - \sqrt{\mathcal{R}}\, o_2^\dagger\right)|00\rangle_{\text{out}}, \\
&= \left[\sqrt{\mathcal{R}(1-\mathcal{R})}\, o_1^\dagger o_1^\dagger - \mathcal{R} o_1^\dagger o_2^\dagger + (1-\mathcal{R}) o_2^\dagger o_1^\dagger - \sqrt{\mathcal{R}(1-\mathcal{R})}\, o_2^\dagger o_2^\dagger\right]|00\rangle_{\text{out}}, \\
&= \left[\sqrt{\mathcal{R}(1-\mathcal{R})}\, o_1^\dagger o_1^\dagger + (1-2\mathcal{R}) o_1^\dagger o_2^\dagger - \sqrt{\mathcal{R}(1-\mathcal{R})}\, o_2^\dagger o_2^\dagger\right]|00\rangle_{\text{out}}, \\
&= \sqrt{2\mathcal{R}(1-\mathcal{R})}\,(|20\rangle_{\text{out}} - |02\rangle_{\text{out}}) + (1-2\mathcal{R})|11\rangle_{\text{out}}.
\end{aligned}
\tag{7.1}
$$

In the final line, we have made use of the ladder operator $a^\dagger|1\rangle = \sqrt{2}|2\rangle$. For the 50:50 beam splitter $\mathcal{R} = 0.5$ this gives

$$
|\psi\rangle_{\text{out}} = \frac{1}{\sqrt{2}}(|20\rangle_{\text{out}} - |02\rangle_{\text{out}}). \tag{7.2}
$$

This is the same remarkable result that we saw in figure (7.1). The photons always leave the beam splitter together. For a detector that does not distinguish photon number, the output is a superposition of photons, no-photons (in the upper, lower channels, respectively), and no-photons, photons. This is an *entanglement state*, equivalent to the so called **Bell state** $|\Psi^-\rangle$, that we shall consider below. The perfect destructive interference in equation (7.1) depends on the photons having identical spatial modes, and the beam splitter being exactly 50:50. If the photons are not identical then the single photon output in each channel is only partially destroyed. For a beautiful experimental demonstration of this partial cancellation, see [1].

We shall illustrate two-photon interference using two figures, figures 7.4 and 7.5 (interactive versions available online). In the first example, figures 7.4, we show the effect of varying the reflectivity, $\mathcal{R}$, of the beam splitter. We consider the case, described in equation (7.1), where there is one photon incident in each input channel. We use the size of the dots to represent the amplitude of the state, with two dots indicating the two photons in the same channel. Colour is used to represent the sign of each term in the Fock state superposition (red for positive, olive for negative). We can make a few interesting observations on the results, in particular, about how two-photon interference differs from one. First, consider the term corresponding to one photon in each output port, the $|11\rangle_{\text{out}}$ term. The probabilities to detect one photon in either output channel are always equal to one another. This is very different to the single-photon input case where the probabilities to detect one photon in one or the other output channel are, in general, unequal. The amplitude of the $|11\rangle_{\text{out}}$ term does depend on the reflectivity. It disappears completely for a 50:50 beam splitter, and changes sign for $\mathcal{R} > 0.5$. This sign change can be significant if this path subsequently interferes with another path.

In the second example shown in figure 7.5 (interactive version available online) we illustrate the effect of varying the input state for the special case of a 50:50 beam splitter. In the static version of the figure, each row depicts a different input state. The columns illustrate propagation in time. A sign change occurs when the photon is

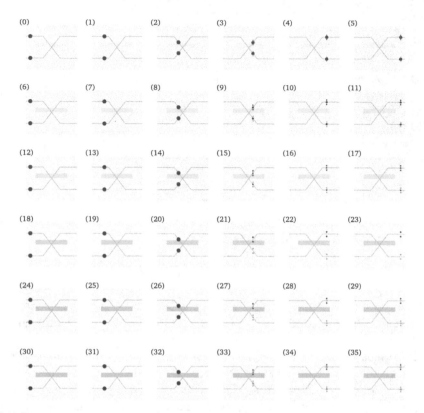

Figure 7.4. Two-photon interference—the HOM effect: the input state in the Fock basis is $|\psi\rangle_{in} = |11\rangle_{in}$. The colomns correspond to different times as the photon propagates through the network. The rows correspond to six different values of the reflectivity $\mathcal{R} = (0.1, 0.25, 0.33, 0.5, 0.67, 0.75)$. An interactive figure is available at http://doi.org/10.1088/978-0-7503-2628-5.

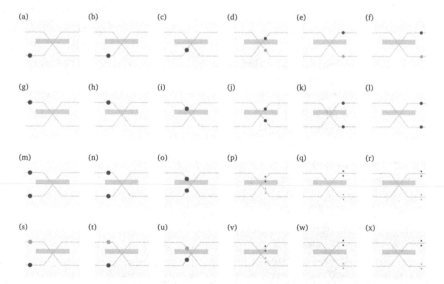

Figure 7.5. Two-photon interference—the Hong–Ou–Mandel (HOM) effect: in the Fock basis $\{n_1, n_2\}$, where n_1 is the upper channel in this plot. (a)–(f) $|1, 0\rangle$, (g)–(l) $|0, 1\rangle$, (m)–(r) $|1, 1\rangle$, and (s)–(x) $|1, -1\rangle$. An interactive figure is available at http://doi.org/10.1088/978-0-7503-2628-5.

reflected at an interface to a medium with higher refractive index. The sign change is represented by a change in colour (similar to the colour wheel concept introduced in the context of the density matrix). The size of the dots corresponds to the amplitude of the state. The two-photon Fock states are represented by two dots. This figure highlights another interesting property of this two-photon interference effect. Apart from a global phase factor, changing the phase of either input photon does not change the output state. This is shown for a $|1, -1\rangle$ input in figures 7.5(s)–(x).

7.4 Bell states

In the previous section, we saw how two-photon interference results in quantum entanglement. The output of the HOM effect, equation (7.2) is a particular example of a special class of states known as **Bell states**—named after **John Bell**. The state vectors of the four Bell states are,

$$|\Phi^{\pm}\rangle = \frac{1}{\sqrt{2}}(|00\rangle \pm |11\rangle),$$

$$|\Psi^{\pm}\rangle = \frac{1}{\sqrt{2}}(|01\rangle \pm |10\rangle).$$

If we do not resolve photon number then equation (7.2) is the same as $|\Psi^{-}\rangle$. The Bell states are special because they form the complete set of maximally entangled two-qubit states. Maximally entangled means one cannot factorize them into separable parts, at all! An important consequence of this impossibility of factorization means that the Bloch sphere picture, where we represent each quantum object separately, no longer works.

7.5 Polarization-entangled photons

The two-photon interference example described above is an example of entanglement in the dual-rail basis. We can also have Bell states in the polarization basis. Polarization-entangled photons arise in two-photon decay of excited states, and parametric down conversion in non-linear optics. Such polarization-entangled photons are hugely significant in the development of quantum optics and our understanding of quantum mechanics. The **2022 Nobel Prize in Physics** was awarded for experiments on polarization-entangled photons.

The first experiments on polarization-entangled photon pairs were performed using two-photon decays from calcium atoms in an atomic beam, see Kocher and Cummins 1967 in the timeline figure in chapter 2. Subsequently, in 1972, Freedman and Clauser showed how the resulting polarization entanglement violated classical counting statistics. In 1981 Aspect, Grangier and Roger demonstrated entanglement survives at long distance[1]. Clauser and Aspect were awarded the Nobel Prize for their contributions in 2022.

A two-photon decay is illustrated in figure 7.6. A Ca atom decays from a state with both electrons excited 4p4p (top) to the ground state 4s4s state (bottom) via

---

[1] For more historical background we recommend this interview with Alain Aspect from 2022 [2].

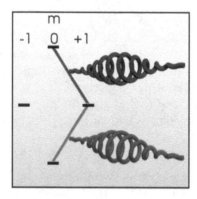

Figure 7.6. Two-photon emssion: for each decay path via the $m = \pm 1$ states (only one shown) two photons are emitted with correlated photon polarizations.

states with only one electron excited 4s4p or 4p4s (middle) via the emission of two photons. The two-photon decay can follow one of two paths—only one is shown. Angular momentum conservation requires that if the photons are emitted in the same direction, the polarizations for each step of the decay are orthogonal. For the path shown, the first (red) and second (blue) photons are right- and left-circularly polarized, respectively. For the other path these polarizations are reversed. Left- and right-circularly polarized photons are emitted when the component of angular momentum of the electron in the direction of emission increases ($\Delta m = +1$, red) or decreases ($\Delta m = -1$, blue) by one unit. As momentum is conserved it follows that the first photon (red) has $m_{\mathrm{s}} = -\hbar$ and is right-circularly polarized, whereas the second photon (blue) has $m_{\mathrm{s}} = +\hbar$ and is left circular.

In the experiment the two photons are detected in opposite directions, which means that they must be either both left- or both right-circularly polarized. We can write this as

$$|\Phi^+\rangle = \frac{1}{\sqrt{2}}(|\circlearrowleft\circlearrowleft\rangle + |\circlearrowright\circlearrowright\rangle) = \frac{1}{\sqrt{2}}(|LL\rangle + |RR\rangle).$$

We can rewrite the state in a linear basis using the relationship between linear and circular polarizations,

$$|L\rangle = \frac{1}{\sqrt{2}}(|H\rangle + i|V\rangle), \quad \text{and} \quad |R\rangle = \frac{1}{\sqrt{2}}(|H\rangle - i|V\rangle).$$

It follows that

$$\frac{1}{\sqrt{2}}(|LL\rangle + |RR\rangle) = \frac{1}{\sqrt{2}}(|HH\rangle - |VV\rangle). \tag{7.3}$$

This is still a Bell state! This transformation highlights one of the bizarre properties of Bell states. We can rotate our measurement basis and still get the same answer. In this example, if the two observers, A and B, measure in the circular basis, then if A measures left, then B measures left, or if A measures right, then B measures right.

Similarly, if they measure in the linear basis, if A measures horizontal then B measures horizontal, or if A measures vertical then B measures vertical. As long as we they choose the same basis, regardless of what basis that is, they get the same answer. But if they chose different bases then they only have a 50% chance of getting the same result. The fact that quantum mechanics can display correlations that go beyond what can be predicted classically is encapsulated in the form of **Bell's inequality**.

7.6 Bell's inequality

In 1964 John Bell showed that joint measurement outcomes on entangled particles cannot be predicted from their local uncertainties. Bell formulated this property in terms of an inequality that can be tested experimentally (figure 7.7). A good introductory discussion of Bell inequalities is given in [3]. Measurement outcomes on a Bell state do not follow from classical probability. Not only are the measurement outcomes correlated but, as we saw in the previous section, these correlations can still appear even when we change the measurement basis. In this section we explore how this strange property of quantum correlation arises due to the basis invariance of Bell states.

Consider two observers A and B each measuring the polarization of one photon in a Bell pair. To calculate the probability that A observes a $|+\rangle$ (which could be a particular linear or circular polarization), we construct an operator that projects photon A onto $|+\rangle$ and does nothing to B[2],

Figure 7.7. Bell inequalities student lab experiment. To see annotated story about details of the experimental setup, including non-linear crystal operation, alignment and phase compensation, scan QR code or go to https://www.researchx3d.com/?o=6&s=11 in order to open experimental setup. An annotated model is based on the setup in the Laboratoire d'Enseignement Expérimental (LEnsE) in the Institut d'Optique Graduate School, Universite Paris-Saclay. [Annotated story opens in web browser online; optionally, R3D marker can also be used to see experiment in Augmented Reality].

---

[2] For a single qubit, a projection operator finds the component of the Bloch vector along a particular direction. Note also for mixed states we have to use density matrices rather than state vectors. In this case, the expectation values are found using a trace,

$$\langle P \rangle = \mathrm{Tr}(\rho P).$$

$$P = |+\rangle\langle+| \otimes \hat{\sigma}_0, = \frac{1}{2}\begin{pmatrix} 1 & 1 \\ 1 & 1 \end{pmatrix} \otimes \begin{pmatrix} 1 & 0 \\ 0 & 1 \end{pmatrix}$$

$$= \frac{1}{2}\begin{pmatrix} 1 & 0 & 1 & 0 \\ 0 & 1 & 0 & 1 \\ 1 & 0 & 1 & 0 \\ 0 & 1 & 0 & 1 \end{pmatrix}.$$

The probability that we detect A in $|+\rangle$ regardless of B is given by the expectation value of this operator

$$\langle\Phi^+|P|\Phi^+\rangle = \frac{1}{4}(1\ \ 0\ \ 0\ \ 1)\begin{pmatrix} 1 & 0 & 1 & 0 \\ 0 & 1 & 0 & 1 \\ 1 & 0 & 1 & 0 \\ 0 & 1 & 0 & 1 \end{pmatrix}\begin{pmatrix} 1 \\ 0 \\ 0 \\ 1 \end{pmatrix},$$

$$= \frac{1}{4}(1\ \ 0\ \ 0\ \ 1)\begin{pmatrix} 1 \\ 1 \\ 1 \\ 1 \end{pmatrix} = \frac{1}{2}.$$

The photon observed by A has a 50% probability to be detected in state $|+\rangle$. Similarly, for B.

Now we could ask, in what direction does the Bloch vector for photon A (or B) point? Surely, there must be some direction (θ, ϕ), where we find that the probability to be detected in that direction is unity? We can attempt to find the direction (θ, ϕ) by using an operator that projects onto the arbitrary angle (θ, ϕ), i.e.

$$P_{|\theta,\phi\rangle} = |\theta,\ \phi\rangle\langle\theta,\ \phi|\otimes\hat{\sigma}_0$$

$$= \frac{1}{2}\begin{pmatrix} 1 + \cos\theta & e^{-i\phi}\sin\theta \\ e^{i\phi}\sin\theta & 1 - \cos\theta \end{pmatrix} \otimes \begin{pmatrix} 1 & 0 \\ 0 & 1 \end{pmatrix}$$

$$= \frac{1}{2}\begin{pmatrix} 1 + \cos\theta & 0 & e^{-i\phi}\sin\theta & 0 \\ 0 & 1 + \cos\theta & 0 & e^{-i\phi}\sin\theta \\ e^{i\phi}\sin\theta & 0 & 1 - \cos\theta & 0 \\ 0 & e^{i\phi}\sin\theta & 0 & 1 - \cos\theta \end{pmatrix}.$$

The expectation value is

$$\langle\Phi^+|P_{|\theta,\phi\rangle}|\Phi^+\rangle = \frac{1}{4}(1\ \ 0\ \ 0\ \ 1)\begin{pmatrix} 1 + \cos\theta \\ e^{-i\phi}\sin\theta \\ e^{i\phi}\sin\theta \\ 1 - \cos\theta \end{pmatrix} = \frac{1}{2}.$$

But this is strange. It says that the probability for the Bloch vector to point is any direction is always one-half. Hence it does not point in any particular direction. It is equally probable to detect it in any direction. Consequently, we can no longer draw a meaningful Bloch sphere for either A or B. Although A and B might be separate in real space we cannot separate their quantum states.

If the probability is to detect A in $|+\rangle$ and B in $|+\rangle$ then classically we expect that the probability to detect both in $|+\rangle$ is $P_{AB}^{++} = P_A^+ P_B^+ = \frac{1}{4}$. But we know from the state vector that $P_{AB}^{++} = \frac{1}{2}$, because they are correlated. To summarize this unspecified direction in the Bloch sphere we can write that:

If we prepare the polarizations of two photons in a Bell state, then the polarization of each photon is undefined, only the correlation between them is defined.

This strange directionless aspect of the Bloch vector is related to rotational invariance that we saw in equation (7.3). In the next section, we expand on this rotational invariance and introduce a visualization based on the density matrix.

7.7 Two-qubit visualization: rotations

Before leaving Bell states we shall consider how to represent them visually—given that the Bloch (or Poincaré) sphere is no longer useful. We shall illustrate our visualization using the experiment of two-qubit rotations.

Single-qubit rotations performed independently on each qubit are given by the tensor product of the single-qubit rotation operators. For example, consider an R_y rotation applied to both qubits. The two-qubit rotation matrix is

$$R_y^{(2)}(\Theta) = R_y(\Theta) \otimes R_y(\Theta),$$

$$= \begin{pmatrix} \cos\dfrac{\Theta}{2} & -\sin\dfrac{\Theta}{2} \\ \sin\dfrac{\Theta}{2} & \cos\dfrac{\Theta}{2} \end{pmatrix} \otimes \begin{pmatrix} \cos\dfrac{\Theta}{2} & -\sin\dfrac{\Theta}{2} \\ \sin\dfrac{\Theta}{2} & \cos\dfrac{\Theta}{2} \end{pmatrix},$$

$$= \frac{1}{2}\begin{pmatrix} 1+\cos\Theta & -\sin\Theta & -\sin\Theta & 1-\cos\Theta \\ \sin\Theta & 1+\cos\Theta & -1+\cos\Theta & -\sin\Theta \\ \sin\Theta & -1+\cos\Theta & 1+\cos\Theta & -\sin\Theta \\ 1-\cos\Theta & \sin\Theta & \sin\Theta & 1+\cos\Theta \end{pmatrix}.$$

If we apply this rotation to the Bell state, $|\Phi^+\rangle$, we find

$$R_y^{(2)}(\Theta)|\Phi^+\rangle = \frac{1}{2}\begin{pmatrix} 1+\cos\Theta & -\sin\Theta & -\sin\Theta & 1-\cos\Theta \\ \sin\Theta & 1+\cos\Theta & -1+\cos\Theta & -\sin\Theta \\ \sin\Theta & -1+\cos\Theta & 1+\cos\Theta & -\sin\Theta \\ 1-\cos\Theta & \sin\Theta & \sin\Theta & 1+\cos\Theta \end{pmatrix}\frac{1}{\sqrt{2}}\begin{pmatrix} 1 \\ 0 \\ 0 \\ 1 \end{pmatrix},$$

$$= \frac{1}{\sqrt{2}}\begin{pmatrix} 1 \\ 0 \\ 0 \\ 1 \end{pmatrix}.$$

Hence the Bell state, $|\Phi^+\rangle$, is invariant to a $R_y^{(2)}(\Theta)$ rotation.

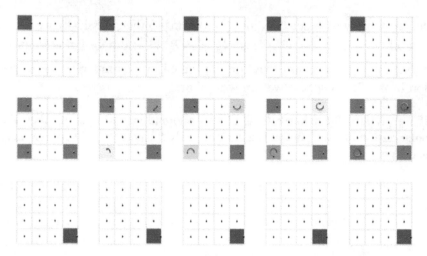

Figure 7.8. Density matrix evolution for a two-qubit $R_z^\Theta \otimes R_z^\Theta$ rotation. Rows 1–3 correspond to input states: $|00\rangle$, a Bell state, $|\Phi^+\rangle = \frac{1}{\sqrt{2}}(|00\rangle + |11\rangle)$, and $|11\rangle$, respectively. For this case only the phases of the $|00\rangle$ and $|11\rangle$ change. The phase change is indicated by the colour and dot which gives both the amplitude and phase in the complex plane. An interactive figure is available at http://doi.org/10.1088/978-0-7503-2628-5.

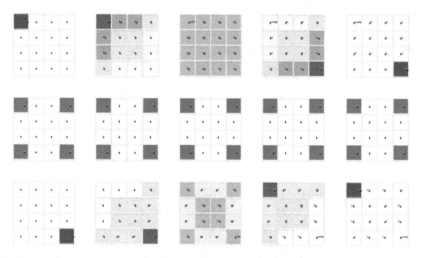

Figure 7.9. R_y rotating acting on both qubits. For the Bell state, $|\Phi^+\rangle$, (middle row) we see the rotational invariance. An interactive figure is available at http://doi.org/10.1088/978-0-7503-2628-5.

In figures 7.8–7.10 we illustrate the effect of different rotations, $R_z^{(2)}(\Theta)$, $R_y^{(2)}(\Theta)$, and $R_x^{(2)}(\Theta)$. For each case we have three input states for different input states, $|00\rangle$, the Bell state,

$$|\Phi^+\rangle = \frac{1}{\sqrt{2}}(|00\rangle + |11\rangle),$$

and $|11\rangle$. The *rotational invariance* of the Bell state appears in the middle row of figures 7.9.

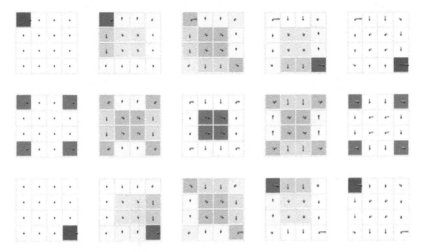

Figure 7.10. R_x rotating acting on both qubits. For the Bell state, $|\Phi^+\rangle$ (middle row), the two-qubit state oscillates between $|\Phi^+\rangle$ and $|\Psi^+\rangle$ (middle column in static figure). An interactive figure is available at http://doi.org/10.1088/978-0-7503-2628-5.

In the following section, we shall consider how it is possible to use the entangling property of two-photon interference to realize a two-qubit quantum gate and hence in principle build a photonic quantum computer.

7.8 Linear optics quantum computing

If we can use two-photon interference to produce entanglement then it follows that we must be able to realize two-photon gates. The combinations of the single-qubit rotations implemented using beam splitters and a two-photon gate is sufficient to create a universal quantum computer. This concept is exploited in **linear optics quantum computing (LOQC)** proposed by Knill, Laflamme, and Milburn (KLM) in 2001 [4]. The KLM scheme uses only beam splitters. Single-qubit rotations are performed as described in Part I. Two-qubit gates make use of two-photon interference. In this section we shall describe one method to realize a controlled-NOT (CNOT) gate. The scheme makes use of both interferometry and two-photon interference (i.e. the HOM effect). As a first step next towards modelling a two-qubit gate we consider a slightly more complex photonic network that differs from the HOM beam splitter in that it is not symmetric.

7.9 Hong–Ou–Mandel effect: non-symmetric network

In this section, we consider an example where $\mathcal{M}^{-1} \neq \mathcal{M}$. The simplest case is where we add a second beam splitter in one path, as in figure 7.11. Although this example is relatively simple with only two photons, two beam splitters, and three channels, it illustrates how complexity grows exponentially on the number of channels or number of photons. In fact, photonic circuits exhibit exactly the same exponential scaling that is the basis of the power of quantum computing.

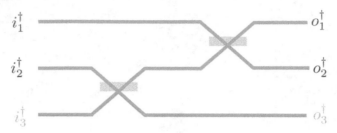

Figure 7.11. Asymmetric network.

Using the beam splitter input–output relations we can write

$$o_1^\dagger = \frac{1}{\sqrt{2}}\left[i_1^\dagger + \frac{1}{\sqrt{2}}(i_2^\dagger + i_3^\dagger)\right]$$

$$o_2^\dagger = \frac{1}{\sqrt{2}}\left[i_1^\dagger - \frac{1}{\sqrt{2}}(i_2^\dagger + i_3^\dagger)\right],$$

$$o_3^\dagger = \frac{1}{\sqrt{2}}(i_2^\dagger - i_3^\dagger),$$

or in matrix form

$$\begin{pmatrix} o_1^\dagger \\ o_2^\dagger \\ o_3^\dagger \end{pmatrix} = \frac{1}{\sqrt{2}} \underbrace{\begin{pmatrix} 1 & \frac{1}{\sqrt{2}} & \frac{1}{\sqrt{2}} \\ 1 & -\frac{1}{\sqrt{2}} & -\frac{1}{\sqrt{2}} \\ 0 & 1 & -1 \end{pmatrix}}_{\equiv \mathcal{M}} \begin{pmatrix} i_1^\dagger \\ i_2^\dagger \\ i_3^\dagger \end{pmatrix}$$

The inverse is

$$\mathcal{M}^{-1} = \frac{1}{\sqrt{2}}\begin{pmatrix} 1 & 1 & 0 \\ \frac{1}{\sqrt{2}} & -\frac{1}{\sqrt{2}} & 1 \\ \frac{1}{\sqrt{2}} & -\frac{1}{\sqrt{2}} & -1 \end{pmatrix}.$$

As an example, consider the input state $|\psi\rangle_{in} = i_1^\dagger i_2^\dagger |000\rangle_{in}$. The output is

$$|\psi\rangle_{out} = \left(\sum_{\mu=1}^{3}\mathcal{M}_{1\mu}^{-1}o_\mu^\dagger\right)\left(\sum_{\nu=1}^{3}\mathcal{M}_{2\nu}^{-1}o_\nu^\dagger\right)|000\rangle_{out},$$

$$= \frac{1}{\sqrt{2}}(o_1^\dagger + o_2^\dagger)\frac{1}{\sqrt{2}}\left[\frac{1}{\sqrt{2}}(o_1^\dagger - o_2^\dagger) + o_3^\dagger\right]|000\rangle_{out},$$

$$= \left[\frac{1}{2\sqrt{2}}(o_1^{\dagger 2} - o_2^{\dagger 2}) + \frac{1}{2}(o_1^\dagger o_3^\dagger + o_2^\dagger o_3^\dagger)\right]|000\rangle_{out},$$

$$= \frac{1}{2}|200\rangle_{out} - \frac{1}{2}|020\rangle_{out} + \frac{1}{2}|101\rangle_{out} + \frac{1}{2}|011\rangle_{out}.$$

One interesting aspect of this result is the sign of the last term. This term has one photon in channel 2 and has a plus sign. However, without the extra photon in channel 1, we would have had

$$|\psi\rangle_{\text{out}} = \left(\sum_{\nu=1}^{3} \mathcal{M}_{2\nu}^{-1} o_\nu^\dagger \right) |000\rangle_{\text{out}},$$

$$= \frac{1}{\sqrt{2}} \left[\frac{1}{\sqrt{2}} (o_1^\dagger - o_2^\dagger) + o_3^\dagger \right] |000\rangle_{\text{out}},$$

$$= \frac{1}{2} |100\rangle_{\text{out}} - \frac{1}{2} |010\rangle_{\text{out}} + \frac{1}{\sqrt{2}} |001\rangle_{\text{out}}.$$

Now the term with a photon in channel 2, $|010\rangle_{\text{out}}$, has a minus sign. This example illustrates that one consequence of two-photon interference may be to change the sign of the single-photon component in a particular channel. This property is particularly useful in certain applications such as the case of linear optics quantum computing as we consider next.

7.10 Controlled-NOT gate

A circuit layout for a LOQC two-qubit controlled NOT (CNOT) gate is shown in figure 7.12. It consists of six waveguides and five beam splitters (labelled 1–5). The two outer channels are auxillaries used to measure whether the gate has been successful. The operators $x_0^\dagger$ and $x_1^\dagger$ create a photon in the upper and lower auxillary channels, respectively. The central four channels encode the control and target photons using dual-rail encoding. The operators $c_0^\dagger$ and $c_1^\dagger$ create a photon in the control '0' and '1' channels, respectively. The operators $t_0^\dagger$ and $t_1^\dagger$ create a photon in the target '0' and '1' channels, respectively.

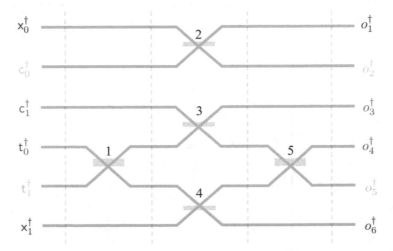

Figure 7.12. Example photonic circuit used in the linear optics quantum computing (LOQC) CNOT gate. The wave guide layout consists of six channels and five beam splitters. Beam splitters 1 and 2 which form the interferometer have a reflectivity of 0.5. Beam splitters 3, 4 and 5 have a reflectivity of 0.33.

In the target channels, there is an interferometer formed by beam splitters 1 and 5, see figure 7.12. Both of these beam splitters are 50:50, forming a Mach–Zehnder identical to figure 6.2. The extra ingredient is that one arm of the interferometer is mixed with control '1' channel on beam splitter 3. All the quantum magic arising from two-photon interference happens at beam splitter 3. But before we discuss the magic we should note that there is a downside to this scheme. By mixing the control and target channels on beam splitter we allow the target photon to be lost to the control channel, or similarly the control photon may be lost to the target channel. To balance these losses, the control '0' and target '1' channels are mixed with auxillary channels on beam splitters 2 and 4. This introduces further losses. Only when one photon is detected in one of the control channels and one photon is detected in one of the target channels does the gate work. Consequently, the gate is probabilistic and only works for a fraction of the attempts—one in nine, as we shall see. This probabilistic nature of the gate makes scaling to large-scale computation extremely challenging. As the KLM scheme is fundamentally lossy, it is not easy to scale LOQC up to a large number of qubits.

To work out the state in different regions of the circuit we use the same matrix formalism described in the previous sections. As there are six channels we have a 6×6 matrix. The matrix describing the first beam splitter is

$$
\mathcal{M}_1 = \begin{pmatrix}
1 & 0 & 0 & 0 & 0 & 0 \\
0 & 1 & 0 & 0 & 0 & 0 \\
0 & 0 & 1 & 0 & 0 & 0 \\
0 & 0 & 0 & \dfrac{1}{\sqrt{2}} & \dfrac{1}{\sqrt{2}} & 0 \\
0 & 0 & 0 & \dfrac{1}{\sqrt{2}} & -\dfrac{1}{\sqrt{2}} & 0 \\
0 & 0 & 0 & 0 & 0 & 1
\end{pmatrix}
$$

Beam splitters 2, 3, and 4 have a reflectivity of one third. The matrix for beam splitters 2–4 is

$$
\mathcal{M}_{2-4} = \begin{pmatrix}
\dfrac{1}{\sqrt{3}} & \dfrac{\sqrt{2}}{\sqrt{3}} & 0 & 0 & 0 & 0 \\
\dfrac{\sqrt{2}}{\sqrt{3}} & -\dfrac{1}{\sqrt{3}} & 0 & 0 & 0 & 0 \\
0 & 0 & \dfrac{1}{\sqrt{3}} & \dfrac{\sqrt{2}}{\sqrt{3}} & 0 & 0 \\
0 & 0 & \dfrac{\sqrt{2}}{\sqrt{3}} & -\dfrac{1}{\sqrt{3}} & 0 & 0 \\
0 & 0 & 0 & 0 & -\dfrac{1}{\sqrt{3}} & \dfrac{\sqrt{2}}{\sqrt{3}} \\
0 & 0 & 0 & 0 & \dfrac{\sqrt{2}}{\sqrt{3}} & \dfrac{1}{\sqrt{3}}
\end{pmatrix}.
$$

Note that for beam splitter 4 it is the upper channel that is reflected from at an interface to higher index and picks up the minus sign.

By multiplying $\mathcal{M}_1$ and $\mathcal{M}_{2-4}$, we obtain the matrix for beam splitters 1–4 is

$$\mathcal{M}_{1-4} = \begin{pmatrix} \dfrac{1}{\sqrt{3}} & \dfrac{\sqrt{2}}{\sqrt{3}} & 0 & 0 & 0 & 0 \\[8pt] \dfrac{\sqrt{2}}{\sqrt{3}} & -\dfrac{1}{\sqrt{3}} & 0 & 0 & 0 & 0 \\[8pt] 0 & 0 & \dfrac{1}{\sqrt{3}} & \dfrac{1}{\sqrt{3}} & \dfrac{1}{\sqrt{3}} & 0 \\[8pt] 0 & 0 & \dfrac{\sqrt{2}}{\sqrt{3}} & -\dfrac{1}{\sqrt{6}} & -\dfrac{1}{\sqrt{6}} & 0 \\[8pt] 0 & 0 & 0 & -\dfrac{1}{\sqrt{6}} & \dfrac{1}{\sqrt{6}} & \dfrac{\sqrt{2}}{\sqrt{3}} \\[8pt] 0 & 0 & 0 & \dfrac{1}{\sqrt{3}} & -\dfrac{1}{\sqrt{3}} & \dfrac{1}{\sqrt{3}} \end{pmatrix}$$

Note that like in the previous section this matrix is non-symmetric. The matrix for the complex network $\mathcal{M}_{1-5} = \mathcal{M}_1\mathcal{M}_{2-4}$,

$$\mathcal{M}_{1-5} = \begin{pmatrix} \dfrac{1}{\sqrt{3}} & \dfrac{\sqrt{2}}{\sqrt{3}} & 0 & 0 & 0 & 0 \\[8pt] \dfrac{\sqrt{2}}{\sqrt{3}} & -\dfrac{1}{\sqrt{3}} & 0 & 0 & 0 & 0 \\[8pt] 0 & 0 & \dfrac{1}{\sqrt{3}} & \dfrac{1}{\sqrt{3}} & \dfrac{1}{\sqrt{3}} & 0 \\[8pt] 0 & 0 & \dfrac{1}{\sqrt{3}} & -\dfrac{1}{\sqrt{3}} & 0 & \dfrac{1}{\sqrt{3}} \\[8pt] 0 & 0 & \dfrac{1}{\sqrt{3}} & 0 & -\dfrac{1}{\sqrt{3}} & -\dfrac{1}{\sqrt{3}} \\[8pt] 0 & 0 & 0 & \dfrac{1}{\sqrt{3}} & -\dfrac{1}{\sqrt{3}} & \dfrac{1}{\sqrt{3}} \end{pmatrix}.$$

Using these matrices we can calculate the state at any position in the circuit for any input state. In the qubit basis, a general two-qubit input state

$$|\psi\rangle = \alpha|00\rangle + \beta|01\rangle + \gamma|10\rangle + \delta|11\rangle.$$

In the Fock basis, this is written as

$$|\psi\rangle_{\text{in}} = (\alpha c_0^\dagger t_0^\dagger + \beta c_0^\dagger t_1^\dagger + \gamma c_1^\dagger t_0^\dagger + \delta c_1^\dagger t_1^\dagger)|000000\rangle$$
$$= \alpha|010100\rangle + \beta|010010\rangle + \gamma|001100\rangle + \delta|001010\rangle$$

We use the operator transform matrix, $\mathcal{M}$, to rewrite the input operators in terms of output operators. As before, to avoid confusion we use new labelling, as shown in figure 7.12. For the $|CT\rangle = |10\rangle$ term, using the transfer matrix $\mathcal{M}_{1-5}$, the inner operator product $\gamma c_1^\dagger t_0^\dagger$ term is transformed into an output product

$$|\psi\rangle_{\text{out}} = \left[\ldots + \gamma \frac{1}{\sqrt{3}}(o_3^\dagger + o_4^\dagger + o_5^\dagger) \frac{1}{\sqrt{3}}(o_3^\dagger - o_4^\dagger + o_6^\dagger) + \ldots \right] |000000\rangle.$$

There are nine possible outcomes but as the $o_3^\dagger o_4^\dagger$ terms cancel only seven terms survive. Writing this in terms of photon numbers, we have

$$|\psi\rangle_{\text{out}} = \frac{1}{3}\left[\ldots + \gamma(|002000\rangle_{\text{out}} - |001100\rangle_{\text{out}} + |001001\rangle_{\text{out}} \right.$$
$$+ |001100\rangle_{\text{out}} - |000200\rangle_{\text{out}} + |000101\rangle_{\text{out}}$$
$$\left. + |001010\rangle_{\text{out}} - |000110\rangle_{\text{out}} + |000011\rangle_{\text{out}}) + \ldots \right]$$
$$= \ldots + \frac{1}{3}\gamma(|002000\rangle_{\text{out}} + |001001\rangle_{\text{out}} - |000200\rangle_{\text{out}}$$
$$+ |000101\rangle_{\text{out}} + |001010\rangle_{\text{out}} - |000110\rangle_{\text{out}} + |000011\rangle_{\text{out}}) + \ldots ,$$

Only one term (the fifth $|001010\rangle_{\text{out}}$) has one photon in both the control and target channels. This corresponds to $|CT\rangle = |11\rangle$ which is the NOT operation we require. So if we post-select on cases where we detect one photon in the control channels and one photon in the target channels, we obtain the result that a $|10\rangle$ input is switched into $|11\rangle$ output, as required. The drawback is that this outcome only occurs with a probability of $(\frac{1}{3})^2 = \frac{1}{9}$. This high failure rate makes scaling of LOQC extremely challenging.

Similarly, for a $|CT\rangle = |11\rangle$ input, the $\delta c_1^\dagger t_1^\dagger$ term transforms into an output state,

$$|\psi\rangle_{\text{out}} = \left[\ldots + \delta \frac{1}{\sqrt{3}}(o_3^\dagger + o_4^\dagger + o_5^\dagger) \frac{1}{\sqrt{3}}(o_3^\dagger - o_5^\dagger - o_6^\dagger) \right] |000000\rangle_{\text{out}}.$$

This time the $o_3^\dagger o_5$ terms cancel leaving only $o_3^\dagger o_4^\dagger$ which again is a NOT operation we require. The output Fock state is

$$|\psi\rangle_{\text{out}} = \ldots + \frac{1}{3}\delta(|002000\rangle_{\text{out}} - |001001\rangle_{\text{out}} + |001100\rangle_{\text{out}}$$
$$- |000110\rangle_{\text{out}} - |000100\rangle_{\text{out}} - |000020\rangle_{\text{out}} - |000011\rangle_{\text{out}}).$$

As before, seven terms survive but the only one with one photon in both the control and target channels is $|001100\rangle$ which corresponds to $|CT\rangle = |10\rangle$. So a $|11\rangle$ input is changed into a $|10\rangle$ output, conditional on detection at least one photon in the control channels and one in the target channels. Again with a success probability

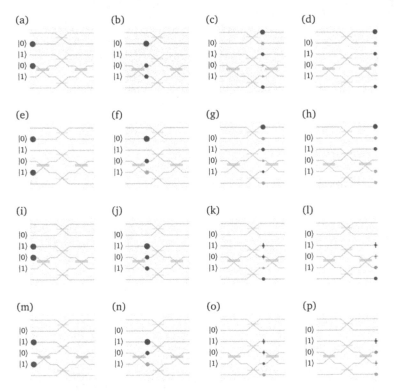

Figure 7.13. Linear optics quantum computing (LOQC) CNOT gate. The four rows correspond to the four input states: in the {C, T} basis, (a)–(d) |00⟩, (e)–(h) |01⟩, (i)–(l) |10⟩, and (m)–(p) |11⟩. An interactive figure is available at http://doi.org/10.1088/978-0-7503-2628-5.

of $\frac{1}{9}$. All four input states are depicted schematically in figure 7.13, where again we use marker size to indicate amplitude and colour to indicate the sign. However, in some cases the sign is ambiguous because there are terms with a photon in a particular channel that have opposite signs that do not cancel. Consequently, this figure does not tell the whole story. The entangling nature of the circuit is hidden in the correlations between channels. This is a good example of the quote we used earlier, if we can see it, then it is not quantum.

To tell the complete story we need to represent all the possible paths in the Fock state superposition. This is illustrated in figure 7.14. We have separated the terms in the superposition using a horizontal offset. For example, if we look at column two in the static figure then the effect of the first beam splitter is to split both the target and the control into parts representing the control plus a target 0 photon (in channel 4) and the control plus a target 1 photon (in channel 5). As we encounter subsequent beam splitters, the dots representing photons are further divided. For example in (c) and (h) we have eight terms and hence eight columns of pair of dots. In the final column we show the show the outcome post-selected of having one photon in the control output channels and one photon in the target photon channels.

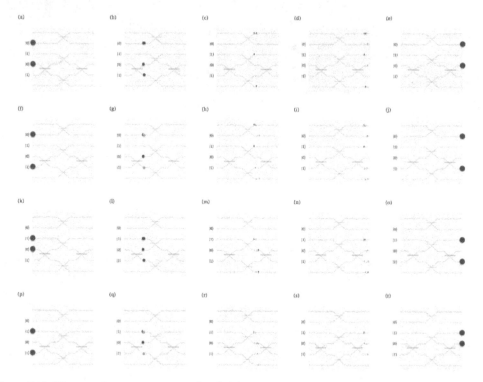

Figure 7.14. Linear optics quantum computing (LOQC) CNOT gate. Similar to figure 7.13 except that we represent each term in the Fock basis superposition using separate filled circles separated along the horizontal axis. In additon, we add an additional final column where we post-select the results that have one photon in each of the control and target channels. These post-selection cases produce the desired CNOT operator. An interactive figure is available at http://doi.org/10.1088/978-0-7503-2628-5.

References

[1] Specht H P *et al* 2009 Phase shaping of single-photon wave packets *Nat. Photon.* **3** 469–72
[2] Phillips W D and Dalibard J 2023 Experimental tests of Bell's inequalities: a first-hand account by Alain Aspect *Eur. Phys. J. D* **77** 8
[3] Maccone L 2013 A simple proof of Bell's inequality *Am. J. Phys.* **81** 854
[4] Knill E, Laflamme R and Milburn G 2001 A scheme for efficient quantum computation with linear optics *Nature* **409** 46–52

IOP Publishing

An Interactive Guide to Quantum Optics

Nikola Šibalić and C Stuart Adams

Chapter 8

Seeing entanglement: counting and correlation

8.1 Introduction

In this chapter we review a particular experiment known as the **quantum eraser**, plus a fascinating variant of this experiment known as the *delayed-choice quantum eraser*. The latter builds on **Wheeler's delayed-choice experiment**, where at some future time we decide whether we want to measure wave- or particle-like properties. We shall devote some time to this experiment because it provides a beautiful exemplar of the concepts of entanglement, counting, correlation and measurement. It also exploits the rotational invariance of the Bell state that we discussed in the previous chapter. The actual experiment was performed by Kim *et al* in 2000[1]. The name is controversial because, in fact, as it can be argued that nothing is, in fact, erased nor delayed [1]. However, the experiment does provide a useful framework to explore concepts relating to wave-particle duality and complementarity, plus the even more exotic questions relating to decoherence and retrocausality—the ability to rewrite the past. Roland Omnès in his book, *The Interpretation of Quantum Mechanics* [2], provides an entertaining discussion of the retrocausality concept,

> An intergalactic civilisation with very powerful technical [...] have read Bell and they make surreptitiously a measurement of the type he proposed. [...] a man, in the year 1000, might have died of a cancer initiated by a unique irradiation originating from the decay of a uranium nucleus [...] He survived [...] and died only thirty years later [...] In the meantime, he had a child who was among the ancestors of Napoleon and also of Doctor Babbit, who now teaches quantum mechanics. The measurement made by the intergalactic jokers [...] contains a component of the nucleus decay products such that the old man died in the year 1000 [...]. The extraterrestials show their results on TV news and Mister Babbit is asked to explain them. What will he say? Necessarily that there

---

[1] See the experiment by Kim *et al* [3].

is no doubt that he himself exists and does not exist, that many books on history are both right and wrong. In a nutshell: facts don't exist as facts and truth is an empty notion.

What is missing here is that unless the Bell pair is isolated from the environment it decoheres and the correlations are destroyed. As Omnès goes on to say *one cannot circumvent decoherence* any more than we can circumvent the second law of thermodynamics!

8.2 Quantum erasure

The basic idea of the delayed-choice quantum erasure is to produce an entangled photon pair, and subsequently decide whether to measure wave- or particle-like properties. According to **complementarity** we cannot see both at the same time. First, we shall look again at the basis invariance of a Bell state in the context of polarized photon pairs. For example, if we take the state

$$|\Phi^+\rangle_z = \frac{1}{\sqrt{2}}(|00\rangle + |11\rangle),$$

this is identical to

$$|\Phi^+\rangle_x = \frac{1}{\sqrt{2}}(|++\rangle + |--\rangle),$$

where $|\pm\rangle = \frac{1}{\sqrt{2}}(|0\rangle \pm |1\rangle)$. This degree of freedom allows an observer, or observers, to decide on their preferred basis. For example, they may chose to measure either in the $\{|0\rangle, |1\rangle\}$ basis or the $\{|+\rangle, |-\rangle\}$ basis.

For two photons, using polarization encoding, we can have a Bell state of the form

$$|\Phi^+\rangle_x = \frac{1}{\sqrt{2}}(|HH\rangle + |VV\rangle), \tag{8.1}$$

where $|H\rangle$ and $|V\rangle$ represent horizontal and vertical polarizations, respectively. This type of state is produced in two-photon decay of, for example, calcium atoms, see section 7.5, or parametric down conversion. Using $|\nearrow\rangle = \frac{1}{\sqrt{2}}(|H\rangle + |V\rangle)$ and $|\searrow\rangle = \frac{1}{\sqrt{2}}(|H\rangle - |V\rangle),$[2] we can also have

$$|\Phi^+\rangle_x = \frac{1}{\sqrt{2}}(|\nearrow\nearrow\rangle + |\searrow\searrow\rangle).$$

---

[2] Note that the global phases of polarization states differ from the standard $\{|0\rangle, |1\rangle\}$ basis used in quantum computing. The states $|0\rangle$ and $|1\rangle$ correspond to left- and right-circular polarization, $|L\rangle$ and $|R\rangle$, respectively. The state $|+\rangle = \frac{1}{\sqrt{2}}(|L\rangle + |R\rangle)$ is equivalent to horizontal polarization state $|H\rangle = \frac{1}{\sqrt{2}}(|L\rangle + |R\rangle)$. However, the state $|-\rangle = \frac{1}{\sqrt{2}}(|L\rangle - |R\rangle)$, where vertical polarization $|V\rangle = -i\frac{1}{\sqrt{2}}(|L\rangle + |R\rangle)$. The extra phase factor arises because the $|L\rangle$ and $|R\rangle$ components are defined as aligned along the horizontal axis at $t = 0$, and we need to advance time to $t = \pi/2\omega$ for them to line up along the vertical axis giving a phase factor $e^{-i\omega t} = e^{-i\pi/2} = -i$. Similarly, the $|+i\rangle = \frac{1}{\sqrt{2}}(|0\rangle + i|1\rangle)$ and $|-i\rangle = \frac{1}{\sqrt{2}}(|0\rangle - i|1\rangle)$ are not identical to the diagonal states $|D\rangle = |\nearrow\rangle = \frac{1}{\sqrt{2}}(e^{-i\pi/4}|L\rangle + e^{i\pi/4}|R\rangle)$ and $|A\rangle = |\searrow\rangle = \frac{1}{\sqrt{2}}(e^{i\pi/4}|L\rangle + e^{-i\pi/4}|R\rangle)$. As the Bloch sphere representation ignores the global phase, these differences are not apparent.

Two observers—often called Alice and Bob—may chose to measure either in the horizontal, vertical basis $\{|H\rangle, |V\rangle\}$, or in the diagonal basis $\{|\nearrow\rangle, |\searrow\rangle\}$. If they choose the same basis their measurements agree, if Alice observes a $|H\rangle$, then Bob will also observe a $|H\rangle$. However, if they choose different bases their results are uncorrelated. If Alice observes a $|H\rangle$ (or a $|V\rangle$) then Bob may still observe either a $|\nearrow\rangle$ or a $|\searrow\rangle$, as apparent if we rewrite equation (8.1) in the form

$$|\Phi^+\rangle_x = \frac{1}{\sqrt{2}}\left[\left|H\rangle\frac{1}{\sqrt{2}}(|\nearrow\rangle + |\searrow\rangle) + \left|V\rangle\frac{1}{\sqrt{2}}(|\nearrow\rangle - |\searrow\rangle)\right.\right].$$

These correlations are used in one possible scheme **quantum cryptography**—the ability to share a secret key using a single photon channel that in principle is immune to an eavesdropper[3].

8.3 Photon experiments

To illustrate these concepts we start with the simple case of single-photon interference using polarized light. Figure 8.1 shows a Mach–Zehnder interferometer where a beam of photons is first split into two paths, and then those two paths are overlapped such that we observe inteference fringes in the overlap region. The experiment can be performed one photon at a time, similar to G I Taylor's variation on Young's double slit (see the timeline figure in chapter 2). To convert this into a 'which-path' experiment we make the first beam splitter a polarizing beam splitter (PBS), figure 8.1(b), where the horizontal (p-) polarization is transmitted and the vertical (s-) polarization is reflected. In this case, the overlapping beams have orthogonal polarizations and we do not observe any interference fringes. However, it is wrong to say that the two beams do not interfere. It's just that the interference pattern is hidden. This accords with complementarity where we would say that we

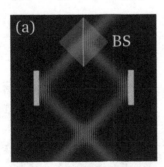 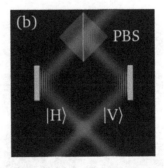

Figure 8.1. (a) A Mach–Zehnder type interferometer where light is split using a beam splitter (BS) into two paths. The two paths are subsequently recombined forming an interference pattern. The horizontal position of the fringes depends on the relative phase between the two paths. (b) The beam splitter is replaced with a PBS which splits the beam into components with horizontal ($|H\rangle$) and vertical ($|V\rangle$) polarizations. In this case no fringes are observed when the two beams overlap.

---

[3] In practice for one-way communication, a single photon is sufficient for quantum cryptography.

can either have a wave-like property—interference—or particle-like properties—which-path information—but not both at the same time. Wheeler's **delayed-choice** experiment involves making the decision about whether to measure particle (path) or wave-like (interference) properties after the photon is 'split' at the beam splitter. Our observation is independent of when we make this choice. This has been verified experimentally[4].

As highlighted, the absence of fringes at the intersection between the horizontally and vertically polarized beams in figure 8.1(b) does not arise due to a lack of interference. Rather, by tracing over the polarization decay of freedom we are summing over the interference pattern in such a way that the fringes are not visible. If we think of the photon as a polarization-encoded qubit, this is equivalent to the sum $|a|^2 + |b|^2$.

To recover the interference pattern it is sufficient to insert a linear polarizer in the overlap region. This is illustrated in figure 8.2. The four rows show the polarizer; horizontal ($\alpha = 0$), along the positive diagonal ($\alpha = 45°$), vertical ($\alpha = 90°$), and along the negative diagonal ($\alpha = 135°$). The columns show what happens when we displace the observation point along in the horizontal direction. In this case the horizontal ($|H\rangle$) and vertical ($|V\rangle$) components transverse different path lengths and acquire a relative phase. We can express this by writing a state vector in the overlap region in the form

$$|\psi\rangle = \frac{1}{\sqrt{2}}(e^{i\phi/2}|H\rangle + e^{-i\phi/2}|V\rangle), \tag{8.2}$$

where ϕ is the phase difference between the two paths. Re-writing this in the diagonal basis, we have

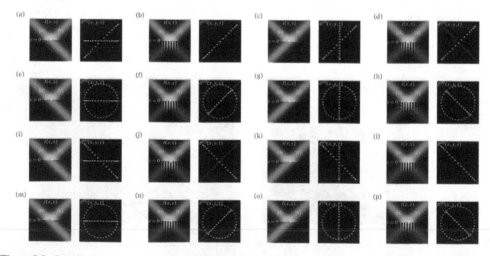

Figure 8.2. Detail of the beam overlap region from with the addition of a linear polarizer. The projection of the field onto the polarizer axis as a function of time is shown in the right in each case.

---

[4] See [4]. In the experiment, single photon pulses enter a 50 m long interferometer. In the time that the photon tranverses the interference, the measurement is switched electronically between which-path or interference.

$$|\psi\rangle = \frac{1}{2}[e^{i\phi/2}(|\nearrow\rangle + |\searrow\rangle) + e^{-i\phi/2}(|\nearrow\rangle - |\searrow\rangle)],$$

$$= \cos\phi/2|\nearrow\rangle + i\sin\phi/2|\searrow\rangle.$$

From this expression, it follows that if we measure $|\nearrow\rangle$ we see cosine^2 fringes and if we measure $|\searrow\rangle$ we see sine^2 fringes and if we measure both we see no fringes. This is illustrated in figures 8.2(b) and (d). In (b) we see cosine^2 fringes, the intensity at the centre marked by the white square is a maximum. In (d) we see sine^2 fringes, the intensity at the centre is minimum.

The accompanying figures on the right show the electric field (red dots), and its projection onto the polarizer axis (blue dots) illustrate how the relative phase and polarizer angle combine to give the observed intensity. For example, in the first column we see that the electric field at the centre is linearly polarized along the positive diagonal, figure 8.2(a), but as we move off centre, it changes to circular, figure 8.2(e), then back to linear along the negative diagonal figure 8.2(i), and then to the opposite circular figure 8.2(m). However, in all cases the projection on the horizontal axis, the red dots in the right-hand plot, is the same. Consequently, we do not see any change in the light intensity as we move along the horizontal axis—no fringes. In constrast, when the polarizer is diagonal, as in figures 8.2(b), (f), (j) and (n), the projection does change with position and the interference pattern becomes visible.

By measuring on the diagonal, i.e. measuring in the $\{|\nearrow\rangle, |\searrow\rangle\}$ basis which are superpositions of $|H\rangle$ and $|V\rangle$, we recover an interference pattern. We could say that the particle-like 'which-path information'—encoded in $|H\rangle$ and $|V\rangle$—has been erased, allowing us to see wave-like interference. This experiment has been called a *single-photon quantum eraser* in the literature, see e.g. [5]. Alternatively, we could explain our observations using classical polarization optics[5].

The experiment becomes more quantum if the single-photon entering the interferometer is a part of a Bell pair. Consider the case where Alice performs an interference measurement and Bob only measures polarization. Alice's photon picks up the relative phase factors as in equation (8.2) and the two-photon state vector is

$$|\Psi\rangle = e^{i\phi/2}|HH\rangle + e^{-i\phi/2}|VV\rangle. \tag{8.3}$$

Re-writing this in the diagonal basis we get

$$|\Psi\rangle = \cos\frac{\phi}{2}(|DD\rangle + |AA\rangle) + i\sin\frac{\phi}{2}(|DA\rangle + |AD\rangle), \tag{8.4}$$

where $|D\rangle \equiv |\nearrow\rangle$ and $|A\rangle \equiv |\searrow\rangle$. Now Alice—who measures the first (red coded) photon—can only observe cosine^2 fringes if both Alice and Bob use the diagonal basis and post-select only the data where they both measure a $|D\rangle$ or both measure an $|A\rangle$. If Alice measures, say a $|D\rangle$, but does not post-select using the results of Bob then both the $|DD\rangle$ and $|AD\rangle$ components are summed giving an intensity proportional to $\sin^2\frac{\phi}{2} + \cos^2\frac{\phi}{2} = 1$. Hence no fringes are observed. In the next section we present an alternative explanation of this effect using quantum information theory.

---

[5] As we do in figure 8.2 or for example figure 4.19 in [6].

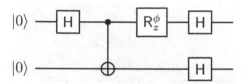

Figure 8.3. The first two layers consisting of a Hadamard gate on qubit A and a CNOT gate create the Bell state. The rotation about z, R_z^ϕ imprints a phase. Finally, two Hadamards acting on both A and B mix the $|0\rangle$ and $|1\rangle$ states.

8.4 A quantum circuit model

The language of quantum computing provides an alternative approach to analysing the quantum eraser without needing to know anything about the polarization of light. A quantum circuit model of the quantum eraser experiment is shown in figure 8.3[6]. The first two layers—consisting of a Hadamard gate on qubit A and a CNOT gate—create the Bell state (we do not need this if as in the optics case we begin with a Bell state). The third layer, R_z^ϕ acting on A, is a rotation around the z axis in the Bloch sphere picture. This imprints a phase such that the Bell state becomes,

$$|\Psi\rangle = \frac{1}{\sqrt{2}}(e^{-i\phi/2}|00\rangle + e^{i\phi/2}|11\rangle).$$

This is the equivalent of equation (8.3). Finally, two Hadamards acting on both A and B mix the $|0\rangle$ and $|1\rangle$ states—equivalent to switching to the diagonal polarization basis.

Using

$$H*H = \frac{1}{\sqrt{2}}\begin{pmatrix} 1 & 1 \\ 1 & -1 \end{pmatrix} * \frac{1}{\sqrt{2}}\begin{pmatrix} 1 & 1 \\ 1 & -1 \end{pmatrix},$$

we obtain the output of the circuit

$$
H*H|\Psi\rangle = \frac{1}{2}\begin{pmatrix} 1 & 1 & 1 & 1 \\ 0 & -1 & 1 & -1 \\ 1 & 1 & -1 & -1 \\ 0 & -1 & -1 & 1 \end{pmatrix} \frac{1}{\sqrt{2}}\begin{pmatrix} e^{-i\phi/2} \\ 0 \\ 0 \\ e^{i\phi/2} \end{pmatrix}
$$

$$
= \frac{1}{\sqrt{2}}\begin{pmatrix} \cos\phi/2 \\ -i\sin\phi/2 \\ -i\sin\phi/2 \\ \cos\phi/2 \end{pmatrix},
$$

---

[6] See also this article: 'Deflating delayed choice quantum erasure' http://algassert.com/quantum/2016/01/07/Delayed-Choice-Quantum-Erasure.html.

which—apart from phase factors that arise due to the differences between the qubit and polarization bases—is the same as equation (8.4). Interference can only be observed if the measurements on A and B are correlated. Measuring either independently gives no interference.

The Hadamard on B is sometimes described as the 'eraser' operation. The idea is that the 'which-path' information about A is encoded in the state of B. If A followed the $|0\rangle$ path, then B remains in $|0\rangle$, whereas if A followed the $|1\rangle$ path then B is switched to $|1\rangle$. By mixing the $|0\rangle$ and $|1\rangle$ using a Hadamard on B, can erase this information? Consider what happens when the Hadamard on B is delayed. Using

$$H*\sigma_0 = \frac{1}{\sqrt{2}}\begin{pmatrix} 1 & 1 \\ 1 & -1 \end{pmatrix} * \begin{pmatrix} 1 & 0 \\ 0 & 1 \end{pmatrix},$$

the state beforehand is

$$H*\sigma_0|\Psi\rangle_3 = \frac{1}{\sqrt{2}}\begin{pmatrix} 1 & 0 & 1 & 0 \\ 0 & 1 & 0 & 1 \\ 1 & 0 & -1 & 0 \\ 0 & 1 & 0 & -1 \end{pmatrix}\frac{1}{\sqrt{2}}\begin{pmatrix} e^{-i\phi/2} \\ 0 \\ 0 \\ e^{i\phi/2} \end{pmatrix}$$

$$= \frac{1}{2}\begin{pmatrix} e^{-i\phi/2} \\ e^{i\phi/2} \\ e^{-i\phi/2} \\ e^{i\phi/2} \end{pmatrix}.$$

Now even if Bob reports a $|0\rangle$ or $|1\rangle$ it does not help recover the interference pattern. But the interference pattern is still there, hidden in the correlations, it is just that if Alice and Bob measure in a difference basis they are effectively chosing not to look at it.

Instead of a measurement in the $|0\rangle, |1\rangle$ basis, Bob can still apply the Hadamard at any time in the future, and report the results such that together with Alice they can reconstruct the interference pattern. The fact that the interference has already taken place does not make any difference, because the ordering of single-photon rotations on different qubits does not change the output state, namely

$$H*H = \left(H*\sigma_0\right)\left(\sigma_0^*H\right) = \left(\sigma_0 \otimes H\right)\left(H*\sigma_0\right),$$

i.e., the effect of two Hadamard simultaneously is the same as first one then the other, or the other way around. What does this mean for Omnès' extraterrestials? They return to Earth to report the results of their measurement on photon B. Does it make a difference? In fact, they cannot change the past. There is no retrocausality. They only find a result consist with the 'measurement' on qubit A even though it was made 1000 years earlier.

References

[1] Kastner R E 2019 Delay choice quantum eraser neither erases nor delays *Found. Phys.* **49** 717
[2] Omnes R 2018 *The Interpretation of Quantum Mechanics* (Princeton, NJ: Princeton University Press)

[3] Kim Y, Yu R, Kulik S, Shih Y and Scully M 2000 Delayed 'choice' quantum eraser *Phys. Rev. Lett.* **84** 1

[4] Jacques V, Wu E, Grosshans F, Treussart F, Grangier P, Aspect A and Roch J-F 2007 Experimental realization of Wheeler's delayed-choice gedanken experiment *Science* **315** 966

[5] Dimitrova T L and Weis A 2010 Single photon quantum erasing: a demonstration experiment *Eur. J. Phys.* **31** 625

[6] Adams C S and Hughes I G 2019 *Optics f2f: From Fourier to Fresnel* 1st edn (oxford University Press) https://doi.org/10.1093/oso/9780198786788.001.0001

Chapter 9

Strong interactions

9.1 Introduction

Photons do not interact. However, photons inside a medium excite electronic oscillations that may interact with other electrons. These interactions map back into effective interactions between photons. A large photon–photon interaction will occur when a single photon modifies the optical response of the medium in a noticeable way. For example, by shifting a resonance by an amount that is a significant fraction of the resonance width. This regime of strong photon interactions is known as **strong coupling**. However, typically at the few photon level, these interactions are too weak to be observable. Consequently, over the last few decades considerable effort has been made to enhance them. The sub-field of strong photon–photon interactions is known as **quantum nonlinear optics**. In the current era, there are various ways to produce strong coupling. The best known example is **cavity QED**, where the photon is strongly-localized on an atom or artificial atom. This scenario is well described by the **Jaynes–Cummings model** that we mentioned briefly in Part I.

The Jaynes–Cummings model is historically significant as it showed how a single atom can be used to detect the quantization of the electromagnetic field. The reason

LabBricks literature graph for this chapter can be found at
https://labbricks.com/#13/96Oxgdcnr7

is that the dynamics of an atom interacting with a quantized field is different to an atom interacting with a classical field. Whereas the classical field has well-define average amplitude giving rise to a single Rabi frequency, a quantized field typically consists of a distribution of different photons number resulting in a spread in Rabi frequency. This spread can be detected via the excitation dynamics of the an atom interacting with the field. The first experimental proof of the qunatization of the field was performed using Rydberg atoms in a cryogenic microwave cavity in 1987 [1]—24 years after Jaynes and Cummings first published their model in 1963. Subsequent experiments by Haroche and others provided the foundations for a new generation of experiment based on **superconducting circuits** —**circuit cavity QED**—where a microwave field interacts with a Josephson loop which plays the role of an artificial atom or qubit. Circuit cavity QED has become one of the leading candidates for quantum computing. Later in this chapter we shall look at some of the exotic photonic-like states that can be produced in these systems.

9.2 Jaynes–Cummings model

The Jaynes–Cummings model[1] adds an additional ingredient to the Rabi model that we used to describe the interaction of a two-level quantum system with an oscillatory electromagnetic field. The extra ingredient is to quantize the field. As the 'atom' or quantum system is already quantized, this next step is often referred to as **second quantization**. The Jaynes–Cummings model is usually applied to the case of a single two-level quantum system in a cavity, as in figure 4.15 from section 4.13. The cavity ensures that the emitter only couples to a single electromagnetic field mode.

A key question is, when is it necessary to quantize the field as well? The answer is that the Jaynes–Cummings model is only necessary if the atom is able to notice the different between zero, one or more photons. This is true if the shift or splitting of the atomic energy levels due to one photon—characterized by the single photon Rabi frequency, g—is larger than the resonance width—characterized by the decay rate of the atom or the field. For an atom in a cavity, this strong coupling regime is given by the inequality $g > \sqrt{\gamma \kappa}$, where γ and κ are the decay rates of the atom and the cavity, respectively. This anharmonic level spacing (leading to a nonlinearity at the single-photon level) is illustrated in figure 9.1. The effect of the interaction is to split the states $|g, n\rangle$, $|e, n-1\rangle$ by $g\sqrt{n+1}$.

The dynamic of the atomic population inversion (for n photons in the cavity) is given by [3]

$$w(t) = \cos\left(2g\sqrt{n+1}\,t\right).$$

This equation states that the atom undergoes Rabi oscillations with Rabi frequency $\Omega_n = 2g\sqrt{n+1}$. Perhaps surprising the Rabi oscillations can still occur even if there are no photons initially, $n = 0$. However, energy conservation (implicit in the light–atom

---

[1] Comparison of quantum and semi-classical radiation theories with application to the beam maser, [2].

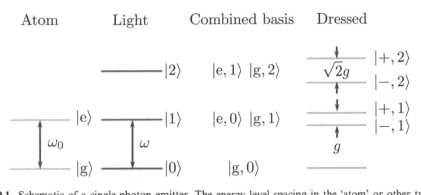

Figure 9.1. Schematic of a single-photon emitter. The energy level spacing in the 'atom' or other two-level system defines the centre frequency. The cavity (blue) and lifetime of the atomic excited define the spatial mode.

interaction term in the Hamiltonian) requires that this can only happen if the atom is initially in the excited state. The exchange of quanta between an excited atom and the cavity mode is known as **vacuum Rabi oscillations**[2]. These oscillations were first observed using Rydberg atoms in a superconducting microwave cavity in 1996 [4]. For a superposition of Fock states with probability P_n, the inversion is given by

$$w(t) = \sum_{n=0}^{\infty} P_n \cos\left(2g\sqrt{n+1}\, t\right).$$

In figure 9.2 we plot this function for a Poissonian distribution of photon numbers. As each photon number produces a different Rabi frequency the population inversion first collapses and then revives.

9.2.1 Exotic multi-quanta states

In principle, we can construct any superposition of Focks state and produce quantum states of light of infinite variety. For example, it is possible to construct pseudo-coherent states which reproduce the Poissonian photon number distribtion of classical light, see e.g. [5]. **However, as these constructed superpositions have well defined phases, they are not the same as 'classical' states.**

Some interesting non-classical cases include **squeezed states**, **cat states**, and **Gottesman–Kitaev–Preskill (GKP) states**. A squeezed state has a Wigner function that is squashed in a particular direction. A cat state has a Wigner function with more than one peak meaning that the electric field is in a superposition of having more than one average value at any time or position. A GKP state has a Wigner function that looks like a grid of lattice.

Experimentally, there are basically two ways to produce exotic quantum states of light, **nonlinearity** or **post-selection**. However, as optical nonlinearities are relatively

---

[2] As the oscillations occur between the states, one ground state atom plus one photon, and one excited atom and no photons, single-photon Rabi oscillations might be a more appropriate descriptor.

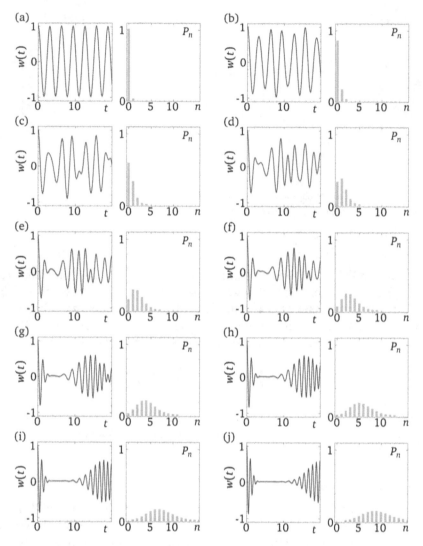

Figure 9.2. Jaynes–Cummings dynamics: the time dependence of the atomic population inversion $z-$ component of the Bloch vector for a single two-level atom interaction with a single mode electromagnetic field with a photon number distribution, P_n, shown in the right hand plot. An interactive figure is available at http://doi.org/10.1088/978-0-7503-2628-5.

small, the scope for manipulating light using nonlinear optics is limited. For this reason some of the most beautiful examples of what might be possible have been demonstrated using circuit cavity QED, where large nonlinearities are accessible in the microwave frequency range. We shall look at some examples below. Post-selection also typically employs nonlinearity but in a different way—to produce correlated photon pairs. By measurement on one partner of a pair—the herald—it is possible to select or engineer the state of the signal photon.

In the context of classical optics, nonlinear means that the light-induced electronic displacement is not a simple linear function of the electric field amplitude. For an oscillatory field, this means that in addition to their harmonic motion, the electrons will also oscillate at harmonics and sub-harmonics of the drive frequency. This nonlinearity can be used to change the frequency of light, for example, either double it which is called second harmonic generation, or halve it which is called parametric down conversion. The latter process is particularly interesting in quantum optics because at the single photon level it means that our single-input photon is converted into two output photons, each with one halve of the energy. Actually, it is possible to split the energy other than 50:50—as long as the process satifies energy and momentum conservation, it is allowed.

9.2.2 Squeezed light

The goal of squeezing is to reduce the noise enabling a more precise measurement. Squeezing is produced using parametric down conversion to create photon pairs and subsequently performing a measurement of components of the pair, see e.g. [6] for an extensive review. Squeezed states are particularly interesting in the context of quantum metrology—because they allow us to beat the shot noise limit of classical measurement Squeezing is widely used to increase the sensity of optical interferometers [7], including gravitational wave detectors [8]. Squeezing is also useful in atom inteferometry [9] and in optical sensing of other fields such as in magnetometry [10].

As Heisenberg's indeterminacy relation says that the product of two quadratures must be greater than $\hbar/2$, it is only possible to reduce the uncertainty in one at the expense of a greater uncertainty in the other. For example, in a mechanical oscillator we can obtain a smaller uncertainty in position at the expense of a larger uncertainty in momentum. For the case of light, we can reduce the uncertainty in photon number at the expense of a larger uncertainty in the phase or vice versa [11]. These two cases are known as amplitude and phase squeezing, respectively.

The electric field for an amplitude squeezed state is shown in figure 9.3. As in figure 4.21, the columns correspond to a different phase of the local oscillator. The row corresponding to increasing the amount of squeezing. Mathematically, squeezed states are characterized by a nonlinear **squeezing operator**,

$$\hat{S}(\xi) = e^{-(\xi \hat{a}^{\dagger 2} + \xi_* \hat{a}^2)},$$

where ξ is the squeezing parameter. Squeezing can arise in an optical medium where the optical response depends nonlinearly on the photon number. However, as noted above, as optical nonlinearities are often small it is more common to employ post-selection.

In a post-selected squeezed state, a nonlinear crystal is employed to produce correlated photon pairs using parametric down conversion. If the two output photons are created in two-modes then the output state can be written in a two-mode photon number basis as

$$|\Psi| = a|00\rangle_0 + b|11\rangle_1 + c|22\rangle_2 + \cdots, \tag{9.1}$$

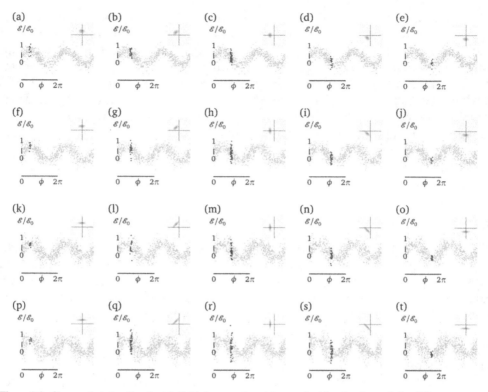

Figure 9.3. Squeezed states: the electric field $\mathcal{E}$ versus time (charaterized by the phase ϕ) for different states of light. Panels (a)–(e) are the vacuum state. Panels (f)–(j) are a single photon state. Panels (k)–(y) show superpositions of photon numbers, n, with intensity P_n indicated by the histograms inset lower right. The corresponding Wigner function is shown as colour map inset upper right. Green is positive and purple is negative. The colums show the time evolution of the state. An interactive figure is available at http://doi.org/10.1088/978-0-7503-2628-5.

where the subscripts refer to the number of photons in the input field. For example, for a single photon input and a 50% probability to down convert then the output is the Bell state Φ_+. In this way we see how nonlinearity and hence an effective interaction between photons produces entanglement. Note that the total photon number of the output is even, although experimentally photon loss in either channel means that the measured photons number can be odd also.

In experiments, it is more common to use a semi-classical input state, namely a coherent state. In this case, the input photon statistics follows a Gaussian distribution with a mean photon number $\bar{n}$ and variance $\sqrt{\bar{n}}$. Down conversion works in the same way on each photon number component resulting in a entangled output state—although the efficiency to create pairs with n-photons each drops off very fast with increasing n. Historically, this bipartite Gaussian entanglement process was called **two-mode squeezing**—as the number of photons in each output mode is correlated, their relative intensity fluctuations are reduced, i.e. **squeezed**. This squeezing property can be use to improve the signal-to-noise and has been used effectively in interferometry such as gravity wave detectors, see [6].

9.2.3 Cat states

The concept of a cat state is that our 'macroscopic' ('classical') thing is in a superposition of being in two states. For a cat, the two states are 'dead' and 'alive'. For light, a cat state is where the electirc field has two or more values at any time. As with squeezing, cat states may be produced via optical nonlinearities or measurement-based post-selection. Let's consider the nonlinear case first, although this effect has only been demonstrated using microwave fields in superconducting circuits[3].

Consider a nonlinear medium where the refractive index depends on the photon number. This type of nonlinearity is called a **Kerr nonlinearity**[4]. In this case, different photon number states acquire different phases as they propagate. For example, if the phase shift per photon is π then a coherent input state evolves into superposition state where the odd and even components are π out of phase. This type of state is often referred to as a cat state, by analogy with Schrödinger's cat, as the field is in a superposition where the expectation value of electric field simultaneously has two values. The non-classicality of a cat state can be detected by constructing the Wigner function, which displays an interference pattern between the two macroscopic values, see figure 9.4. This interference pattern does not occur for a statistical mixture of two fields with a π difference.

A cat state can be also be engineered by selecting only the odd or even photon number states. This has been demonstrated for micorwave fields via a non-destructive parity measurement using Rydberg atoms flying through a cavity, see e.g. [14]. In the case of optics, a cat state has been engineered via a homodyne measurement on a split photon number state, see [15].

We can extend the cat superposition to more than two macroscopic states. Figure 9.5 shows the Wigner function for superpositions of two, three and four coherent states with different phase. A cat state in a superpositon of N values is when the phase shift per photon phase is $2\pi/N$. States of this type have been produced using circuit cavity QED, see e.g. [16].

Finally, in figure 9.6 we compare the Wigner function representation with the density matrix in the photon number basis for particular examples. The three rows show a coherent state, a cat state and an amplitude squeezed state, respectively. As we move across the rows we increase the field amplitude, α. In this case we have chosen a post-selected cat state where only even total photon numbers survive, i.e. $\rho_{mn} = 0$ for $m + n$ odd. This case is different to the Kerr-induced cat state of Vlastaki *et al*, where the odd and even components have opposite sign.

---

[3] There are also beautiful experiments on the motion of ions in harmonic traps that demonstrate the same physics, see e.g. [12, 13].
[4] In the *optical Kerr effect*, the presence of one photon modifies the resonance of the medium such that a second photon experiences a different refractive index. This effective interaction between photons produces entanglement.

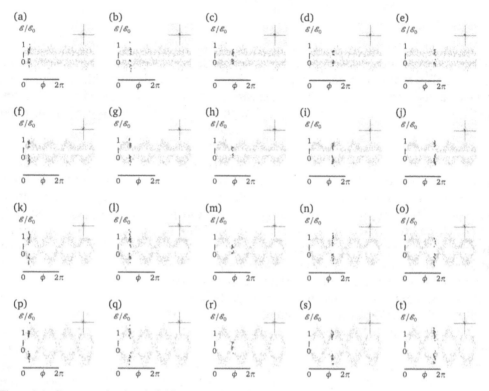

Figure 9.4. Cat states: the electric field $\mathcal{E}$ versus time (charaterised by the phase ϕ) for different states of light. Panels (a)–(e) are the vacuum state. Panels (f)–(j) are a single photon state. Panels (k)–(y) show superpositions of photon numbers, n, with intensity P_n indicated by the histograms inset lower right. The corresponding Wigner function is shown as a colour map inset upper right. Green is positive and purple is negative. The colums show the time evolution of the state. An interactive figure is available at http://doi.org/10.1088/978-0-7503-2628-5.

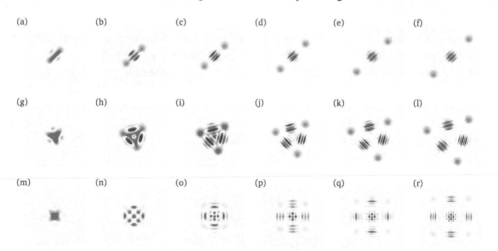

Figure 9.5. Fractional Cat states. The Wigner function for phase shift per photon equal to π (top row), $2\pi/3$ (middle row) and $\pi/2$ (bottom). The column show the effect of increase the mean number of photons. Green is positive and purple is negative. The colums show the time evolution of the state. An interactive figure is available at http://doi.org/10.1088/978-0-7503-2628-5.

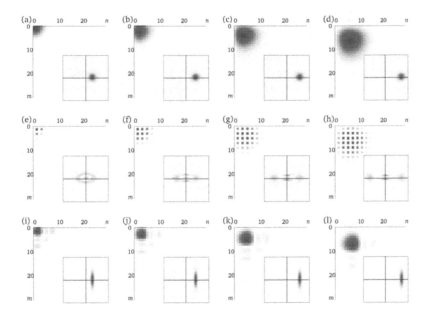

Figure 9.6. The density matrix in the Fock basis, ρ_{mn}, and the corresponding Wigner functions for a coherentn states, a cat state and a squeezed state. An interactive figure is available at http://doi.org/10.1088/978-0-7503-2628-5.

References

[1] Rempe G, Walther H and Klein N 1987 Observation of quantum collapse and revival in a one-atom maser *Phys. Rev. Lett.* **58** 353

[2] Jaynes E T and Cummings F W 1963 Comparison of quantum and semiclassical radiation theories with application to the beam maser *Proc. IEEE* 51 89–90

[3] Narozhny N B, Sanchez-Mondragon J J and Eberly J H 1981 Coherence versus incoherence: collapse and revival in a simple quantum model *Phys. Rev.* A **23** 236

[4] Brune M, Schmidt-Kaler F, Maali A, Dreyer J, Hagley E, Raimond J M and Haroche S 1996 Quantum Rabi oscillation: a direct test of field quantization in a cavity *Phys. Rev. Lett.* **76** 1800

[5] Steindl P 2021 Artificial coherent states of light by multiphoton interference in a single-photon stream *Phys. Rev. Lett.* **126** 143601

[6] Schnabel R 2017 Squeezed states of light and their applications in laser interferometers *Phys. Rep.* **684** 1–51

[7] Caves C M 1980 Quantum-mechanical radiation-pressure fluctuations in an interferometer *Phys. Rev. Lett.* **45** 75

[8] McCuller L *et al* 2021 LIGO's quantum response to squeezed states *Phys. Rev.* D **104** 062006

[9] Hosten O *et al* 2016 Measurement noise 100 times lower than the quantum-projection limit using entangled atoms *Nature* **529** 505

[10] Troullinou C *et al* 2021 Squeezed-light enhancement and backaction evasion in a high sensitivity optically pumped magnetometer *Phys. Rev. Lett.* **127** 193601

[11] Walls D F 1983 Squeezed states of light *Nature* **306** 141–6

[12] Kienzler D *et al* 2016 Observation of quantum interference between separated mechanical oscillator wave packets *Phys. Rev. Lett.* **116** 140402

[13] Flühmann C and Home J P 2020 Direct characteristic-function tomography of quantum states of the trapped-ion motional oscillator *Phys. Rev. Lett.* **125** 043602

[14] Deleglise S *et al* 2008 Reconstruction of non-classical cavity field states with snapshots of their decoherence *Nature* **455** 510

[15] Ourjoumtsev A 2007 Generation of optical Schrödinger cats from photon number states *Nature* **448** 784

[16] Vlastakis B, Kirchmair G, Leghtas Z, Nigg S E, Frunzio L, Girvin S M, Mirrahimi M, Devoret M H and Schoelkopf R J 2013 Deterministically encoding quantum information using 100-photon Schrödinger cat states *Science* **342** 607–10

Part III

Outlook

IOP Publishing

An Interactive Guide to Quantum Optics

Nikola Šibalić and C Stuart Adams

Chapter 10

Outlook

In conclusion, it appears that in all its branches physics is still an experimental science. Its basic goal is not mathematical elegance or the achievement of tenure, but learning the truth about the world around us...
> P W Anderson 1990 Some thoughtful words (not mine)
> on research strategy for theorists *Physics Today* **43** 2

And now for something completely different. Congratulations! You reached this page. Does this complete a quantum optics introduction? Not really, we just had to stop writing at some point. We could have included a chapter on Bell's inequalities, adiabatic following and explained the geometric phase. Or discussed collective excitations, that are so useful when propagating photons interact with some averaged continuous medium. We could also proceed to analyse all sorts of nonlinear optics phenomena based on the fundamental understanding that hopefully we have now. But with the basics established so far you should be able to go and create your own explorable interactive figures on your own[1]. But even that does not lead us to the point where we want to be. For researchers, learning known facts is just the beginning, since our job is to look over the wall on the other side, the unexplored side.

We are leaving teaching how to become expert in dots of knowledge, and want to discuss how one can connect dots and move in the landscape. Physics evolves every day, and what it means to be physicists changes correspondingly. In the present large world we are in dire need of improving knowledge communication and influence decision making on large scales. It is a complex problem without a known solution, yet it is too important to be left out of the discussion. We are taking an orthogonal

---

[1] And please share back what you make!

direction to the proceeding pages, moving out of the plane to better see the previous pages from a new viewpoint—to try and illustrate something more general. In the following we introduce four notes that are our—maybe provocative!—conversation starters.

10.1 Semiotics*

The Universe is made of stories, not atoms.

Muriel Rukeyser

Science is a quest for objectivity, but it is activity that is part of our human existence. And as such, even when we write equations, and decide what is meaningful to group together, and what should be separated, we are influenced by particular stories and viewpoints that we are following at that moment. Based on this we also select some elements for visualization. It is important to keep in mind that there are usually many such projected stories consistent with current data. This has been a hard lesson to learn historically. Switching from an Earth-centric to Helio-centric model didn't go smoothly. More recently, wave and geometric optics pictures had a long history of superseding each other, that ultimately culminated in the present (mixed) quantum view, that is so far consistent with observations.

We shouldn't shy away from trying to imagine new visualizations. Lines of force representing **E** and **B** fields did not appear from nowhere. There was *someone* who started using them—Michael Faraday—and the rest of us followed. Now we learn them early on in our physics education, and we tend to see them as reality—maybe even a bit more than they really are. The little triangles that we use in vector calculus is notation introduced by Oliver Heaviside while he was learning—on his own—Maxwell electromagnetism, written originally in long sets of equations. Feynmann's dots and lines made quantum electrodynamic perturbative calculations so tractable that many more people could do them. The quantum circuit notation of quantum computing allows us to recast an optics problem and gain deeper insight into apparent paradoxical outcomes such as the quantum eraser (as we saw). All of

---

*The study of signs and symbols and of their meaning and use. Its potential power for knowledge transmission was underlined by John Locke (1632–1704) in his '*An Essay Concerning Human Understanding*' first published in 1690: '*Semeiotike, or the doctrine of signs; the most usual whereof being words, it is aptly enough termed also Logike, logic: the business whereof is to consider the nature of signs, the mind makes use of for the understanding of things, or conveying its knowledge to others. For, since the things the mind contemplates are none of them, besides itself, present to the understanding, it is necessary that something else, as a sign or representation of the thing it considers, should be present to it: and these are ideas. And because the scene of ideas that makes one man's thoughts cannot be laid open to the immediate view of another, nor laid up anywhere but in the memory, a no very sure repository: therefore to communicate our thoughts to one another, as well as record them for our own use, signs of our ideas are also necessary: those which men have found most convenient, and therefore generally make use of, are articulate sounds. The consideration, then, of ideas and words as the great instruments of knowledge, makes no despicable part of their contemplation who would take a view of human knowledge in the whole extent of it. And perhaps if they were distinctly weighed, and duly considered, they would afford us another sort of logic and critic, than what we have been hitherto acquainted with.*'

these visualizations—lines of force, vector fields, the Bloch sphere quantum circuits —are our *semiotic* tools that help us understand a world we cannot see.

10.2 Finding and connecting insights

Science is not a soliloquy. It gains value only within its cultural milieu, only by having contact with all those who are now, and who in future will be, engaged in promoting spiritual culture and knowledge. The extant scientific papers of Archimedes, the dialogues and discourses of Galileo, are still of genuine interest in our day, and not only to philologists, but to many scientists. Would it mean setting ourselves too high and too proud a goal, if we occasionally thought of what will have become of our scientific papers 2,000 years hence? Science will have changed entirely. Will there be anybody to grasp our meaning, as we grasp the meaning of Archimedes?

Erwin Schrödinger in Are there quantum jumps?—part II,
The British Journal for the Philosophy of Science **III(11)**, 233 (1952)

Many of the important challenges we face as humanity rely on coordinated work of many experts and organizations. To be able to mobilize a society-wide response to rise to the scale of the challenge, we need to enable knowledge to be not only more easily accessible, but also more easily reused by domain non-experts. We also need more transparency into origins of expert insights, to enable progression when expert opinions differ, and to enable feedback from non-experts that might crucially expand our solution-space horizons. Existing solutions of knowledge dissemination and reuse are not fit for the scale of interdisciplinarity and coordination we have to achieve: we need a new knowledge infrastructure.

If we make good interfaces for finding and reusing knowledge, we can enable society-wide collaboration. Many people pointed in similar directions throughout history, from Vannevar Bush in 'As we may think' (*The Atlantic*, July 1945), over Douglas C Engelbart in 'Augmenting Human Intellect: A Conceptual Framework' (October 1962) to Bret Victor in 'The Humane Representation of Thought' (2014). What is maybe now more evident than ever is that knowledge accessibility is only the first step toward knowledge reuse and application in everyday life. A new knowledge interface has to enable better and easier learning, reuse and communication across the domain and societal boundaries.

Since writing papers still seems the best output for condensed thought we need better tools to organize the ever increasing body of knowledge. The literature record is after all just another source of data that can be analysed and plotted per our needs. For this book we made and used the software LabBricks to generate literature views, connecting ideas and discussions across papers, with text search highlighting links and showing graphically how they track over time. Hopefully in these and similar ways we can share the basis for our conclusions and expertise, allowing them to be challenged, and also providing knowledge 'bricks' more readily usable in other explorations.

10.3 Sharing in accessible and reusable form

Restricting the body of knowledge to a small group deadens the philosophical spirit of a people and leads to spiritual poverty.

Albert Einstein, in foreword to *The Universe and Dr Einstein* by Lincoln Barnett (1948)

How far can we push accessibility and reusability of knowledge? Can we take the letters on the page in the published manuscripts and convert them into code that calculates on demand answers to an infinity of questions? Can we put these codes on the internet to deliver these tools world-wide, and use open-source to actively engage with a diverse community? We tried in this book doing that by publishing as part of Interactive-Publishing GitHub repository Python libraries[2] that we developed for generation of interactive figures, as well as specialized visualizations of quantum states and dynamics that we have developed.

Making something accessible (open data, and open source) does not however mean that it is easily reusable, which should be the goal of the open science movement. Documentation of the interface[3] to your 'knowledge brick' and examples to boot strap new users is crucial for reusability.

Can we make our tools and calculations even more reusable? Imagine that you want to just quickly, over a lunch break, see predictions of global warming under a changed spectrum of irradiation or different atmospheric composition. This should be very easy. We do stand on the shoulders of giants, and many of these models are old. The importance of atmospheric composition for Earth's temperature was underlined by Joseph Fourier in 1824. That CO_2 and H_2O are efficient in trapping heat was shown by John Tyndall in 1860, and by 1896 Svante Arrhenius was estimating temperature variation due to CO_2 changes. Our major problem is not science, it is partly communication, and then what we do with this knowledge.

What we did (not) do with this knowledge in two centuries since Fourier's writing has two problems. The big one is the decision to indulge in blissful ignorance on a societal scale almost up to now. The other one is, however, probably solvable by a small group of dedicated people: we couldn't and *still* cannot take that knowledge and over lunch quickly connect many sources of existing information and models and get predictions. We would need to code, to extract data etc. Does it have to be that way?

The massive boom in applications and speed with which they are developed in the last two decades has a lot to do with the creation of cloud computing, and in particular, creation of reusable primitives (messaging queues, notification systems, databases etc) that developers can quickly mix and match in myriad of forms,

---

[2] **HANDS-ON** ifigures Python package used for creation of interactive figures in this book, together with specialized quantum state visualizations can be installed from Python Package Index (PyPI) online simply by using pip install interactive-publishing.

[3] Also called application programming interface, or API.

corresponding to the plethora of different apps that we have today. If we have models and data as similar primitives accessible online, we could use some (serverless?) framework to connect them and change their parameters and re-run calculations and visualizations. We shouldn't worry about setting up environments, mixing new Python codes and old Fortran codes. We should worry about physics and science more exclusively. As part of this book we try to provide the example of such a service—science web services—allowing online visualizations based on methods developed for this book without any installation[4]. Everyone can copy their state vectors and Hamiltonians, enter them in online forms, and see results. The next step would be to connect such services together.

Finally, why should we share just results? Why just data? Can we share experiments online? After all, many of the top-range experiments take millions in public funding to create, yet the access to them is still mostly limited to a fraction of potentially relevant physics or global communities. We should probably try making them more systematically accessible. The same goes for even simpler educational experiments[5], which are in a large world, still actually a privilege decided by accidents in funding and geographical proximity.

10.4 Science versus decision making

> *Prestige is like a powerful magnet that warps even your beliefs about what you enjoy. It causes you to work not on what you like, but what you'd like to like.*
>
> Paul Graham, from 'How to Do What You Love'

> *Mass hypnosis. In a very orthodox form known as education.*
> Robert M Pirsig in *Zen and the Art of the Motorcycle Maintenance: an inquiry into values*

Science *does not* in general 'direct political decisions' or any decisions for that matter. Deciding is guided by values, and those are the domain of philosophy. Science can provide just the best estimate for all asked questions in the form 'what happens if' and, crucially, *only* to the questions that are asked. *What* is being asked and what value is ascribed to our estimates is again the domain of philosophy. Science therefore tells us about the topography of the possibility landscape, but our motivation for movement in that landscape, the direction we take, and even what we see as space in which we can move is completely open for discussion that has to do with our values, not (exclusively) our science. Even our motivation for particular

---

[4] **HANDS-ON** Access quantum state visualisations online without installation on Science Web Services via https://sws.labbricks.com/qstate.

[5] **HANDS-ON** Currently, you can share and annotate experimental setups and research apparatus designs on ResearchX3D https://www.researchx3d.com. In the future we expect to extend Science Web Services to include live experiments.

research or technological development is determined by our philosophy of values, not the scientific method. This implies that:

- Decision and policy makers cannot hide behind scientific advisors when they are reluctant to make decisions.
- Equally, scientists cannot hide behind pretend objectivity when deciding on research directions and should openly embrace discussions on value systems, knowing that this is not science but is important.
- The intellectual superiority argument 'let the expert decide' has historically been used for suppressing discussions and forcing decisions, leading to discrimination, dogmatism, suffering and damaged democracy. Even experts can be guided by the public to expand their solution space. Politics and philosophy are necessary ingredients for human decision making.
- Agreements cannot be reached just by scientific arguments repeated in the loop. We need to be engaged more often in discussing implied value systems, since shared perception of objective reality is not sufficient for reaching shared vision.

Only the experience of sharing the common human world with others who look at it from different perspectives can enable us to see reality in the round and to develop a shared common sense. Without it, we are each driven back on our own subjective experience, in which only our feelings, wants, and desires have reality.

Margaret Canovan (1939–2018)

These last four sections are not science, but are probably important for science. Their exploration led us to make some of the tools and examples in this book. Maybe they are completely wrong, but probably the best we can do, to escape the traps of ideologies and fashions, is to try to more consciously cultivate a wide range of ideas. By examining, selecting and saving ideas, coming back to them and watering them by rethinking, engaging with others, and weeding out the old opinions, diversity is the best means to ensure growth in the knowledge gardens. Keeping our enthusiasm unrestricted, because it is the psychological fuel that propels us.

The lesson we most need to learn is that there is more to life on Earth than human beings and more to being human than self-interest. Our futures all depend on learning this lesson by heart.

Sir Ken Robinson (2018)

Nikola Šibalić and C Stuart Adams

Appendix A

A.1 Bloch vector

In the z-basis, the Pauli matrices are

$$\sigma_x = \begin{pmatrix} 0 & 1 \\ 1 & 0 \end{pmatrix}, \quad \sigma_y = \begin{pmatrix} 0 & -i \\ i & 0 \end{pmatrix}, \quad \sigma_z = \begin{pmatrix} 1 & 0 \\ 0 & -1 \end{pmatrix},$$

and the identity matrix

$$I_2 \equiv \sigma_0 = \begin{pmatrix} 1 & 0 \\ 0 & 1 \end{pmatrix},$$

where the subscript 2 indicates a 2×2 identity matrix[1]. In quantum computing, it is common to define the Pauli spin matrices as

$$\mathsf{X} = \begin{pmatrix} 0 & 1 \\ 1 & 0 \end{pmatrix}, \quad \mathsf{Y} = \begin{pmatrix} 0 & -i \\ i & 0 \end{pmatrix}, \quad \mathsf{Z} = \begin{pmatrix} 1 & 0 \\ 0 & -1 \end{pmatrix}.$$

and they are often called the **Pauli-X**, **Pauli-Y** and **Pauli-Z**.

Using the Pauli operator we can find u, v, and w as follows:

$$u = \langle \psi | \sigma_x | \psi \rangle$$

$$= \left(\cos\frac{\theta}{2} \quad e^{-i\phi}\sin\frac{\theta}{2} \right) \begin{pmatrix} 0 & 1 \\ 1 & 0 \end{pmatrix} \begin{pmatrix} \cos\frac{\theta}{2} \\ e^{i\phi}\sin\frac{\theta}{2} \end{pmatrix},$$

$$= \left(\cos\frac{\theta}{2} \quad e^{-i\phi}\sin\frac{\theta}{2} \right) \begin{pmatrix} e^{i\phi}\sin\frac{\theta}{2} \\ \cos\frac{\theta}{2} \end{pmatrix},$$

$$= e^{i\phi}\sin\frac{\theta}{2}\cos\frac{\theta}{2} + e^{-i\phi}\sin\frac{\theta}{2}\cos\frac{\theta}{2} = \sin\theta\cos\phi,$$

---

[1] Note that some books include a spin $-1/2$ angular factor $\hbar/2$ in the definition of the spin matrices, i.e. the spin $-1/2$ angular momentum operator is $\boldsymbol{J} = (\hbar/2)\boldsymbol{\sigma}$.

$$v = \langle \psi | \sigma_y | \psi \rangle$$

$$= \left(\cos \frac{\theta}{2} \quad e^{-i\phi} \sin \frac{\theta}{2} \right) \begin{pmatrix} 0 & -i \\ i & 0 \end{pmatrix} \begin{pmatrix} \cos \frac{\theta}{2} \\ e^{i\phi} \sin \frac{\theta}{2} \end{pmatrix},$$

$$= \left(\cos \frac{\theta}{2} \quad e^{-i\phi} \sin \frac{\theta}{2} \right) \begin{pmatrix} -i e^{i\phi} \sin \frac{\theta}{2} \\ i \cos \frac{\theta}{2} \end{pmatrix},$$

$$= -i e^{i\phi} \sin \frac{\theta}{2} \cos \frac{\theta}{2} + -i e^{-i\phi} \sin \frac{\theta}{2} \cos \frac{\theta}{2} = \sin \theta \sin \phi,$$

$$w = \langle \psi | \sigma_z | \psi \rangle,$$

$$= \left(\cos \frac{\theta}{2} \quad e^{-i\phi} \sin \frac{\theta}{2} \right) \begin{pmatrix} 1 & 0 \\ 0 & -1 \end{pmatrix} \begin{pmatrix} e^{i\phi} \sin \frac{\theta}{2} \\ \cos \frac{\theta}{2} \end{pmatrix},$$

$$= \left(\cos \frac{\theta}{2} \quad e^{-i\phi} \sin \frac{\theta}{2} \right) \begin{pmatrix} e^{i\phi} \sin \frac{\theta}{2} \\ \cos \frac{\theta}{2} \end{pmatrix},$$

$$= \cos^2 \frac{\theta}{2} - \sin^2 \frac{\theta}{2} = \cos \theta.$$

A.2 Density matrix

For a pure state described by the state vector $|\psi\rangle$, the density matrix is defined as

$$\rho = |\psi\rangle\langle\psi|. \tag{A.1}$$

If we write the state vector as a column vector

$$|\psi\rangle = \begin{pmatrix} a \\ b \end{pmatrix}, \tag{A.2}$$

then

$$\rho = \begin{pmatrix} a \\ b \end{pmatrix} (a^* \quad b^*) = \begin{pmatrix} |a|^2 & ab^* \\ ba^* & |b|^2 \end{pmatrix}. \tag{A.3}$$

Alternatively written in spherical polar coordinates, the two-state density matrix is

$$\rho = \begin{pmatrix} \cos\dfrac{\theta}{2} \\ e^{i\phi}\sin\dfrac{\theta}{2} \end{pmatrix} \left(\cos\dfrac{\theta}{2} \quad e^{-i\phi}\sin\dfrac{\theta}{2} \right),$$

$$= \begin{pmatrix} \cos^2\dfrac{\theta}{2} & \sin\dfrac{\theta}{2}\cos\dfrac{\theta}{2}e^{-i\phi} \\ \sin\dfrac{\theta}{2}\cos\dfrac{\theta}{2}e^{i\phi} & \sin^2\dfrac{\theta}{2} \end{pmatrix}. \tag{A.4}$$

Using the half-angle formulas, $\sin^2\frac{\theta}{2} = \frac{1}{2}(1 - \cos\theta)$, $\cos^2\frac{\theta}{2} = \frac{1}{2}(1 + \cos\theta)$ and $\sin\frac{\theta}{2}\cos\frac{\theta}{2} = \frac{1}{2}\sin\theta$, and expanding the exponential, we find

$$\rho = \frac{1}{2}\begin{pmatrix} 1 + \cos\theta & \sin\theta\cos\phi - i\sin\theta\sin\phi \\ \sin\theta\cos\phi + i\sin\theta\sin\phi & 1 - \cos\theta \end{pmatrix}.$$

Using the Cartesian components of the Bloch vector,

$$u = \sin\theta\cos\phi,$$
$$v = \sin\theta\sin\phi,$$
$$w = \cos\theta,$$

where $(u^2 + v^2 + w^2)^{1/2} = 1$.

$$\rho = \frac{1}{2}\begin{pmatrix} 1 + w & u - iv \\ u + iv & 1 - w \end{pmatrix}. \tag{A.5}$$

Comparing to the (a, b) form in equation (2), we can also write that

$$u = ab^* + ba^*,$$
$$v = \frac{1}{i}(a^*b - ab^*),$$
$$w = |a|^2 - |b|^2,$$

Note that as

$$ab^* + ba^* = (a_r + ia_i)(b_r - ib_i) + (b_r + ib_i)(a_r - ia_i)$$
$$= a_rb_r + a_ib_i + i(a_ib_r - a_rb_i) + a_rb_r + a_ib_i - i(a_ib_r - a_rb_i)$$
$$= 2(a_rb_r + a_ib_i) = 2\Re[ab^*]$$

and

$$a^*b - ab^* = -2i(a_ib_r - a_rb_i) = -2i\Im[ab^*]$$

both u and v are real as expected (figure A.1).

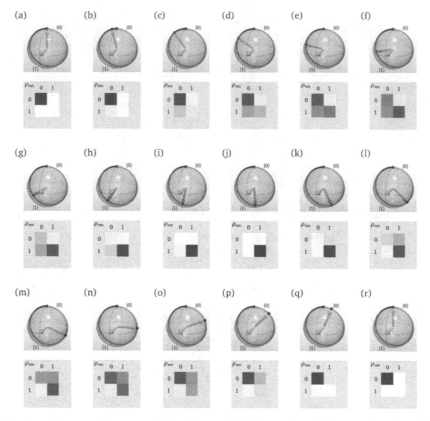

Figure A.1. A rotation about the x axes. The initial state is $|0\rangle$. The rotation axis is shown in yellow. The Bloch vector in blue. The final frame (r) corresponds to a rotation of $\Theta = 2\pi$. Note that the rotation direction follows a right-hand rule. The corresponding density matrix is shown below. An interactive figure is available at http://doi.org/10.1088/978-0-7503-3151-7/978-0-7503-2628-5.

A.3 Rotations

First, we rewrite $\mathsf{R}_{\hat{n}}^{\Theta}$ as a matrix. Using a Taylor expansion and that $\sigma_j^{2m} = \sigma_0$ for any integer m, we have

$$e^{-i\sigma_j(\Theta/2)} = \sigma_0 - i\sigma_j\left(\frac{\Theta}{2}\right) - \frac{\sigma_j^2}{2!}\left(\frac{\Theta}{2}\right)^2$$

$$+ i\frac{\sigma_j^3}{3!}\left(\frac{\Theta}{2}\right)^3 + \frac{\sigma_j^4}{4!}\left(\frac{\Theta}{2}\right)^4 + \cdots$$

$$= \sigma_0 - \frac{\sigma_j^2}{2!}\left(\frac{\Theta}{2}\right)^2 + \frac{\sigma_j^4}{4!}\left(\frac{\Theta}{2}\right)^4 + \cdots$$

$$- i\left[\sigma_j\left(\frac{\Theta}{2}\right) - \frac{\sigma_j^3}{3!}\left(\frac{\Theta}{2}\right)^3 + \cdots\right]$$

$$= \sigma_0 \cos\left(\frac{\Theta}{2}\right) - i\sigma_j \sin\left(\frac{\Theta}{2}\right).$$

Note that in the second line, we rearranged the terms into even and odd series to correspond to cosine and sine, respectively. Using this expansion we can write

$$R_{\hat{n}}^{\Theta} = e^{-i\sigma \cdot \hat{n}(\Theta/2)} = \sigma_0 \cos\left(\frac{\Theta}{2}\right) - i\sigma \cdot \hat{n} \sin\left(\frac{\Theta}{2}\right).$$

Writing the axis of rotation in terms if its component, $\hat{n} = (n_x, n_y, n_z)$, we obtain

$$R_{\hat{n}}^{\Theta} = \begin{bmatrix} \cos\dfrac{\Theta}{2} - in_z \sin\dfrac{\Theta}{2} & (-in_x - n_y)\sin\dfrac{\Theta}{2} \\ (-in_x + n_y)\sin\dfrac{\Theta}{2} & \cos\dfrac{\Theta}{2} + in_z \sin\dfrac{\Theta}{2} \end{bmatrix}. \tag{A.6}$$

The special case of rotation about the x, y, and z axes are given by

$$R_x^{\Theta} = e^{-i(\Theta/2)X} = \cos\frac{\Theta}{2}I - i\sin\frac{\Theta}{2}X$$

$$= \begin{pmatrix} \cos\dfrac{\Theta}{2} & -i\sin\dfrac{\Theta}{2} \\ -i\sin\dfrac{\Theta}{2} & \cos\dfrac{\Theta}{2} \end{pmatrix}.$$

$$R_y^{\Theta} = e^{-i(\Theta/2)Y} = \cos\frac{\Theta}{2}I - i\sin\frac{\Theta}{2}Y$$

$$= \begin{pmatrix} \cos\dfrac{\Theta}{2} & -\sin\dfrac{\Theta}{2} \\ \sin\dfrac{\Theta}{2} & \cos\dfrac{\Theta}{2} \end{pmatrix}.$$

$$R_z^{\Theta} = e^{-i(\Theta/2)Z} = \cos\frac{\Theta}{2}I - i\sin\frac{\Theta}{2}Z$$

$$= \begin{pmatrix} e^{-i\frac{\Theta}{2}} & 0 \\ 0 & e^{i\frac{\Theta}{2}} \end{pmatrix}.$$

For rotations of $\pi/2$ we have the following examples:

$$R_x^{\pi/2}|0\rangle = \frac{1}{\sqrt{2}}\begin{pmatrix} 1 & -i \\ -i & 1 \end{pmatrix}\begin{pmatrix} 1 \\ 0 \end{pmatrix} = \frac{1}{\sqrt{2}}(|0\rangle - i|1\rangle) = |-\rangle_y,$$

$$R_y^{\pi/2}|0\rangle = \frac{1}{\sqrt{2}}\begin{pmatrix} 1 & -1 \\ 1 & 1 \end{pmatrix}\begin{pmatrix} 1 \\ 0 \end{pmatrix} = \frac{1}{\sqrt{2}}(|0\rangle + |1\rangle) = |+\rangle_x,$$

$$R_z^{\pi/2}|+\rangle_x = \begin{pmatrix} e^{-i\frac{\pi}{4}} & 0 \\ 0 & e^{i\frac{\pi}{4}} \end{pmatrix}\frac{1}{\sqrt{2}}\begin{pmatrix} 1 \\ 1 \end{pmatrix} = e^{-i\frac{\pi}{4}}\frac{1}{\sqrt{2}}\begin{pmatrix} 1 \\ e^{i\frac{\pi}{2}} \end{pmatrix}$$

$$= e^{-i\frac{\pi}{4}}|+\rangle_y.$$

The last example illustrates how the Bloch sphere representation is insensitive to the global phase. Rotations around the z and y directions are illustrated in figures A.2–4.10. For $\hat{n} = (\sin\phi, \cos\phi, 0)$ we rotate around a vector in the equatorial plane with azimuthal angle ϕ and

$$R_\phi^\Theta = \begin{pmatrix} \cos\dfrac{\Theta}{2} & -\,\mathrm{i}e^{-\mathrm{i}\phi}\sin\dfrac{\Theta}{2} \\ -\,\mathrm{i}e^{\mathrm{i}\phi}\sin\dfrac{\Theta}{2} & \cos\dfrac{\Theta}{2} \end{pmatrix}.$$

The rotation we found for the 50:50 beam splitter, equation (4.8), is equivalent to a $\pi/2$ rotation about the y axis, as in figure 4.10.

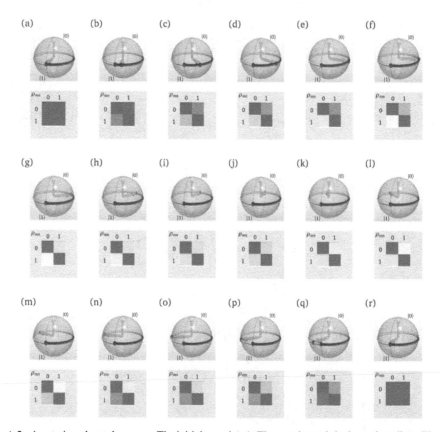

Figure A.2. A rotation about the z axes. The initial state is $|+\rangle$. The rotation axis is shown in yellow. The Bloch vector in blue. The final frame (r) corresponds to a rotation of $\Theta = 2\pi$. Note that the rotation direction follows a right-hand rule. The corresponding density matrix is shown below. An interactive figure is available at http://doi.org/10.1088/978-0-7503-3151-7/978-0-7503-2628-5.

A.4 The Rabi solution

Starting from the Schrödinger equation,

$$i\hbar\partial_t|\psi\rangle = (\mathcal{H}_0 + \mathcal{H}')|\psi\rangle,$$

where $\mathcal{H}_0$ is the Hamiltonian for the quantum system and $\mathcal{H}'$ is the perturbation due to the electromagnetic field. Substituting a state vector of the form

$$|\psi(t)\rangle = a(t)|g\rangle e^{-iE_0 t/\hbar} + b(t)|e\rangle e^{-iE_1 t/\hbar},$$

and using $\mathcal{H}_0|k\rangle = E_k|k\rangle$ to cancel some terms on the left and right gives

$$i\hbar(\dot{a}(t)|0\rangle e^{-iE_0 t/\hbar} + \dot{b}(t)|e\rangle e^{-iE_1 t/\hbar})$$
$$= \mathcal{H}'|0\rangle a(t)e^{-iE_0 t/\hbar} + \mathcal{H}'|e\rangle b(t)e^{-iE_1 t/\hbar}.$$

Next, we take the inner product with $\langle 0|e^{iE_0 t/\hbar}$ (or $\langle e|e^{iE_1 t/\hbar}$), and use the fact that the field couples states $|0\rangle$ and $|1\rangle$, i.e. $\langle 0|\mathcal{H}'|0\rangle = \langle 1|H'|1\rangle = 0$. We are left with

$$i\hbar\dot{a}(t) = \langle 0|\mathcal{H}'|e\rangle e^{-i\omega_0 t}\ b(t),$$
$$i\hbar\dot{b}(t) = \langle e|\mathcal{H}'|0\rangle e^{i\omega_0 t}\ a(t).$$

Typically, the interaction is of the form $\mathcal{H}' = -\boldsymbol{d}\cdot\mathcal{E}$, where $\boldsymbol{d} = -e\boldsymbol{r}$ is the electric dipole operator. If the phase is zero at $t = 0$ then the light is, $\phi_L = \boldsymbol{k}\cdot\boldsymbol{r}$. Note that this depends on the position of the atom which is a problem if the atom is delocalized or moves. We define a **Rabi frequency** (as a measure of the coupling between the field and the qubit) as $\Omega = -\langle e|\boldsymbol{d}\cdot\mathcal{E}_0|0\rangle/\hbar^2$. If the state vectors are real such that $\Omega = -\langle 0|\boldsymbol{d}\cdot\mathcal{E}_0|e\rangle/\hbar$ as well, then we can rewrite the coupled equations as

$$i\dot{a}(t) = \Omega\cos(\phi_L - \omega t)\ e^{-i\omega_0 t}\ b(t)$$
$$i\dot{b}(t) = \Omega\cos(\phi_L - \omega t)\ e^{i\omega_0 t}\ a(t)$$

where $\omega_0 = (E_1 - E_0)/\hbar$. Now we expand the cosine and use that for $\omega \sim \omega_0$, we can neglect terms in $\omega + \omega_0$ (known as **rotating wave approximation**) Integrating $e^{i(\omega+\omega_0)t} + e^{i(\omega-\omega_0)t}$ we get

$$\frac{e^{i(\omega+\omega_0)t}}{i(\omega+\omega_0)} + \frac{e^{i(\omega-\omega_0)t}}{i(\omega-\omega_0)}.$$

For optical fields, typical parameters give $\omega + \omega_0 \sim 10^{15}$ and $\omega - \omega_0 \sim 10^7$, so we can neglect the $\omega + \omega_0$ term, giving

$$i\dot{a}(t) = \frac{1}{2}\Omega e^{-i\phi_L}e^{i\Delta t}\ b(t)$$

$$i\dot{b}(t) = \frac{1}{2}\Omega e^{i\phi_L}e^{-i\Delta t}\ a(t).$$

---

[2] As $\boldsymbol{d} = -e\boldsymbol{r}$ the real frequency is positive.

Finally, we transform into a **rotating frame** by make a change of variable, $\tilde{a} = ae^{-i(\Delta/2)t}$, $\tilde{b} = be^{i(\Delta/2)t}$. This removes the explicit time dependence and we obtain,

$$i\dot{\tilde{a}} = \frac{1}{2}\Delta\tilde{a} + \frac{1}{2}\Omega e^{-i\phi_L}\,\tilde{b}$$

$$i\dot{\tilde{b}} = \frac{1}{2}\Omega e^{i\phi_L}\,\tilde{a} - \frac{1}{2}\Delta\tilde{b}.$$

We can rewrite these coupled equations in the form of a Schrödinger equation,

$$i\hbar\frac{\partial}{\partial t}|\psi\rangle = \mathcal{H}_{\text{int}}|\psi\rangle$$

with

$$\mathcal{H}_{\text{int}} = \frac{\hbar}{2}\begin{pmatrix} \Delta & \Omega e^{-i\phi_L} \\ \Omega e^{i\phi_L} & -\Delta \end{pmatrix},$$

and

$$|\psi\rangle = \begin{pmatrix} \tilde{a} \\ \tilde{b} \end{pmatrix}.$$

We shall omit the tildes from now on. Using the Pauli spin matrices, the atom–light interaction Hamiltonian can be rewritten as

$$\mathcal{H}_{\text{int}} = \frac{\hbar}{2}\left[\Delta\sigma_z + \Omega(\cos\phi_L\sigma_x + \sin\phi_L\sigma_y)\right].$$

Apart from the addition phase factor this is the same as equation (4.12) that we introduced in the context of the classical limit of the Jaynes–Cummings model at the beginning of this chapter. If the interaction Hamiltonian is time-independent then the solution to the Schrödinger equation is

$$|\psi(t)\rangle = \exp(-i\mathcal{H}_{\text{int}}t/\hbar)|\psi(0)\rangle = U|\psi(0)\rangle,$$

where t is the duration of the atom–light interaction. The unitary operator, $U = \exp(-i\mathcal{H}_{\text{int}}t/\hbar)$ has the forms of a two-state rotation matrix. To find the rotation matrix we rewrite the interaction in the form

$$R_{\hat{n}}^{\Theta} = \exp(-i\mathcal{H}_{\text{int}}t/\hbar) = \exp(-i\frac{1}{2}\sigma\cdot\hat{n}\Theta),$$

where Θ is the rotation angle and the unit vector $\hat{n}$ defines the axis of rotation. By comparing the exponentials we find that the unit vector is

$$\hat{n} = \frac{1}{(\Omega^2 + \Delta^2)^{1/2}}[\Omega\cos\phi_L, \Omega\sin\phi_L, \Delta],$$

where the normalization factor $\Omega_{\text{eff}} = (\Omega^2 + \Delta^2)^{1/2}$ is called the effective Rabi frequency. The rotation angle is

$$\Theta = (\Omega^2 + \Delta^2)^{1/2}t.$$

Combining these expressions, we obtain the general rotation matrix

$$R_{\hat{n}}(\Theta) = \begin{bmatrix} \cos\dfrac{\Theta}{2} - i\dfrac{\Delta}{\Omega_{\text{eff}}}\sin\dfrac{\Theta}{2} & -i\dfrac{\Omega}{\Omega_{\text{eff}}}e^{-i\phi_L}\sin\dfrac{\Theta}{2} \\ -i\dfrac{\Omega}{\Omega_{\text{eff}}}e^{i\phi_L}\sin\dfrac{\Theta}{2} & \cos\dfrac{\Theta}{2} + i\dfrac{\Delta}{\Omega_{\text{eff}}}\sin\dfrac{\Theta}{2} \end{bmatrix}.$$

This is known as the *Rabi solution*.

A.5 Quantum regression theorem and two-time correlation calculations

Motivated by discussion in chapter 5.2.1, here we want to calculate a spectrum of a two-level atom driven by strong monochromatic field, and by doing so to demonstrate useful result for general calculation of two- and multi-time expectations of the operators based on quantum regression theorem. We will calculate the spectrum as Fourier transform of the electric field autocorrelation function[3].

Scattered electric field E is proportional to dipole operator $\sigma_\pm$

$$E(\mathbf{r}, t) = \frac{E^+ + E^-}{\sqrt{2}} \tag{A.7}$$

$$E^+(\mathbf{r}, t) = G(\mathbf{r}, \mathbf{d})\sigma_-(t - r/c) \tag{A.8}$$

$$G(\mathbf{r}, \mathbf{d}) = \frac{k^2 e^{ikr}}{4\pi\varepsilon_0 r^3}(\mathbf{d} \times \mathbf{r}) \times \mathbf{r} \tag{A.9}$$

where $\sigma_+ \equiv |e\rangle\langle g|$ and $\sigma_- \equiv |g\rangle\langle e|$, and the last equation is just far field propagation in space $\mathbf{r}$ for a field of a dipole radiator (same as in classical physics) for dipole orientation $\mathbf{d}$. Therefore, to calculate two-time correlation function of the electric field, we can calculate two-time expectation value of atomic dipole operator

$$\langle E^-(t + \tau)E^+(t)\rangle \propto \langle\sigma_+(t + \tau)\sigma_-(t)\rangle. \tag{A.10}$$

This in turn can be calculated using quantum regression theorem

$$\langle\sigma_+(t + \tau)\sigma_-(t)\rangle = \text{Tr}[\sigma_+ V(t, t + \tau)[\sigma_-\rho(t)]], \tag{A.11}$$

where $V(t, t + \tau)$ evolves the density matrix from between two detection events. To find this evolution operator that acts on density matrix, we start with the density matrix master equation for open system in Lindblat form

---

[3] That Fourier transform of the autocorrelation function is equal to the spectrum of that function is the result known as Wiener–Khinchin theorem.

$$\frac{d\rho}{dt} = -i[\mathcal{H}, \rho] + \left(L\rho L^\dagger - \frac{1}{2}L^\dagger L\rho - \frac{1}{2}\rho L^\dagger L\right), \tag{A.12}$$

where dissipation is accounted for with $L \equiv \sqrt{\Gamma}\sigma_-$, and coherent atom driving is given by $\mathcal{H} = \begin{pmatrix} 0 & \Omega/2 \\ \Omega/2 & \Delta \end{pmatrix}$. In the above equations density matrix ρ is a matrix, but we can show the density matrix as *a vector* and represent equation (A.12) as action of a single matrix $\mathcal{L}$ on the vector ρ. For example, in the case of a two-level system, we can write

$$\frac{d}{dt}\begin{pmatrix} \rho_{gg} \\ \rho_{ge} \\ \rho_{eg} \\ \rho_{ee} \end{pmatrix} = \underbrace{\begin{pmatrix} 0 & i\frac{\Omega}{2} & -i\frac{\Omega}{2} & \Gamma \\ i\frac{\Omega}{2} & i\Delta - \frac{\Gamma}{2} & 0 & -i\frac{\Omega}{2} \\ -i\frac{\Omega}{2} & 0 & -i\Delta - \frac{\Gamma}{2} & i\frac{\Omega}{2} \\ 0 & -i\frac{\Omega}{2} & i\frac{\Omega}{2} & -\Gamma \end{pmatrix}}_{\equiv \mathcal{L}}\begin{pmatrix} \rho_{gg} \\ \rho_{ge} \\ \rho_{eg} \\ \rho_{ee} \end{pmatrix}. \tag{A.13}$$

Formally, this can be now solved as

$$\rho(t + \tau) = V(t, t + \tau)[\rho(t)] = e^{\mathcal{L}\tau}\rho(t). \tag{A.14}$$

To find $e^{\mathcal{L}\tau}$ we can decompose $\mathcal{L}$ using right X_i and left Y_i eigenvectors—defined as $\mathcal{L}X_i = \lambda_i X_i$, and $Y_i^\dagger \mathcal{L} = \lambda_i Y_i^\dagger$—to write $\mathcal{L} = \sum_i X_i \lambda_i Y_i^\dagger$. This then allows us to write $e^{\mathcal{L}\tau}$ simply as[4]

$$e^{\mathcal{L}\tau} = \sum_k \frac{\tau^k}{k!}\mathcal{L}^k \tag{A.15}$$

$$= \sum_k \frac{\tau}{k!}\left(\sum_i X_i \lambda^k Y_i^\dagger\right)^k \tag{A.16}$$

$$= \sum_i X_i e^{\lambda_i \tau} Y_i^\dagger \tag{A.17}$$

For a given $\mathcal{L}$ we can numerically find easily[5] right eigenvectors X_i and eigen values λ_i, while Y_i can be found as components of inverse of matrix of right eigenvectors X_i.

---

[4] Below we use the fact that we can always find left eigenvectors Y_i that will be orthogonal with X_i such that $Y_i^\dagger X_i = \delta_{i,j}$.

[5] For example, using in Python programming language numpy.eig method.

This allows calculation of equation (A.14), which can be replaced in equation (A.11) yielding

$$\langle E^-(t + \tau)E^+(t)\rangle \propto \langle \sigma_+(t + \tau)\sigma_-(t)\rangle = \mathrm{Tr}[\sigma_+ e^{\mathcal{L}\tau}\sigma_-\rho(t)] \tag{A.18}$$

where trace Tr[...] is over all system degrees of freedom (in this case $|e\rangle$ and $|g\rangle$). For steady state correlation values, initial t is not crucial, as $\rho(t)$ will be a steady state density matrix, and the result depends only on τ.

Printed in the USA
CPSIA information can be obtained
at www.ICGtesting.com
JSHW060737120224
56719JS00007B/40